Stocking Up

How to Preserve the Foods You Grow, Naturally

By the staff of
Organic Gardening and Farming

Edited by
Carol Stoner

Rodale Press, Inc. Book Division, Emmaus, Pennsylvania 10849

Standard Book Number 0-87857-070-5

Library of Congress Card Number
73-5160

PRINTED IN THE UNITED STATES OF AMERICA

OB-434

EIGHTH PRINTING—DECEMBER 1974

Contents

INTRODUCTION

Since ORGANIC GARDENING AND FARMING's earliest days we have witnessed a growing need for a book that thoroughly covers the subject of natural food preservation. That need is at its peak right now, because not since the victory gardens of World War II have more people been raising and preserving so much of their own food as today. And people are not just raising tomatoes and green beans and strawberries, but a whole variety of vegetables and fruits, nuts, grains, and livestock as well. The number of organic gardeners and homesteaders has grown in recent years because people are discovering that just about the only way they can control the quality of the food they eat is to grow it themselves.

Each year with seasonal regularity, ORGANIC GARDENING AND FARMING's Reader Service Department receives hundreds of letters asking questions about food storage. Spring brings letters requesting building plans for outdoor storage and ideas for converting basements into root cellars. In early summer readers want information on freezing early vegetable varieties and on making jams and jellies with berries and honey. The letter flow builds through the peach and tomato seasons, and by fall, it has developed into a virtual flood: How can I best keep my squash? Should I can or freeze my corn? Do you have a good recipe for apple butter? How do I make sauerkraut using little or no salt? In the winter, readers want to know how to smoke meats and how to make natural cheese from their goat's or cow's milk.

To find the information our readers wanted we first turned to modern books on home food preservation. But we found some problems with the information these books had to offer. Those written in the last thirty years or so contain advanced storage techniques that best preserve the taste, appearance, and nutritive value of the fresh foods, but they are far from complete guides to home food storage. Rather, the modern literature available falls into one of two categories: Either it is books on fruit and vegetable storage for the backyard gardener who wants to can or freeze his garden surplus, or it is booklets and manuals on large-scale crop storage for farmers who have several hundred acres in field crops to hold over the winter months until they are put on the market or fed to livestock. There is pitiful little information for the person who has more than a backyard garden, but less than a big farm operation.

What's more, we didn't feel comfortable advising ORGANIC GARDENING AND FARMING readers to consult any of the books we did find because none of them expressed our concern for organically grown and naturally preserved and prepared foods. ORGANIC GARDENING AND FARMING readers are very particular about their food. Lots of care, thought, and hard work goes into its raising. Organic growers don't take any shortcuts because they know that chemicals that do a quick and easy job of fertilizing soil and speeding up animals' rates of growth may increase the quantity of food produced, but only at a sacrifice in quality, to say nothing of the possible dangers of the chemical residues found in these foods. Organic gardeners and farmers take the time and effort to raise their crops and animals in as natural an environment as possible because they know that the great taste and nutritional value of a tomato or an apple or a chicken that was raised by organic methods is head and shoulders above that of their chemically grown counterparts. These people know that they have high-quality food and they want to keep it that way. They certainly don't want to preserve it with overprocessed or highly refined ingredients or use a method of storage that would unnecessarily destroy any of their food's natural qualities.

In an attempt to find more complete information on natural ways to preserve home-grown food, we looked to books written years ago, when raising and preserving one's own food without the aid of chemical fertilizers, sprays, hormones, medications, and additives was a real part of life for millions of American families. We found that these books are fairly complete guides to home food storage. They provide information on just about everything that could be raised in rural America. Unfortunately, these books are long out-of-print, and although some copies might exist in small public libraries, second-hand bookstores, and in the homes of rare book collectors, they are hard to find.

In addition, a lot of the information in these old books is dated. Old books don't include modern methods of food preservation which have added so much to the ease and safety of home food storage. Hampered by a lack of modern equipment, like steam-pressure canners and home freezers, and a limited knowledge of nutrition, farmers in the past had to rely on methods of food storage that did not always do the best job of retaining the essential nutrients of the foods they were keeping.

We soon realized that if we wanted a book that was a complete guide to food storage—one that had solid information on preserving everything that could be raised on a homestead and one that would express our philosophy about organic foods—we would have to write it ourselves.

And this is just what we did. We have combined what we feel to be the best of the new and the traditional methods of food preservation into a book designed to be a complete guide for organic gardeners, homesteaders, and family farmers who know the satisfaction of raising and preserving their own food as naturally as possible, without the use of any chemicals or over-processed ingredients.

We are grateful to many people for supplying us with information for this book. We would like to thank the Library of Congress for helping us locate both new and long out-of-print, rare books; the U.S. Department of Agriculture and local extension stations for providing us with booklets and pamphlets; and Erewhon Trading Company, Walnut Acres, and other food processors and distributors for the information they have given us on the storage of natural and organically grown foods.

Our most important source of information, however, was our ORGANIC GARDENING AND FARMING readers, because they are the people who are out there raising and harvesting and storing their own food. Our special thanks go to all the organic growers who contributed ideas for food preservation through the articles they have written for our magazine and through the interviews they granted us especially for this book.

VEGETABLES AND FRUITS

CHOOSING VEGETABLE AND FRUIT VARIETIES

As the grower of your own food, you have many advantages over the supermarket shopper. Not only can you choose the fruits and vegetables that you want, but you can also choose the particular varieties of fruits and vegetables that best suit you.

If you page through any seed catalog you'll discover that each food is usually available in a number of varieties. Some of these varieties may be particularly good for freezing; others maintain their quality best when canned. Other varieties have been designed for drying, and some hold their flavor and texture well in underground storage. If you're planning to preserve a good part of your harvest, you'd do well to decide how you will store your garden surplus before you order your seeds, and then choose those fruit and vegetable varieties developed specifically for your method of storage.

Some of the more popular fruits and vegetables, like tomatoes, sweet potatoes, lima beans, beets, cabbage, carrots, potatoes, cauliflower, apples, blueberries, and strawberries are available in varieties particularly high in vitamin content. Although no kind of preservation will increase the vitamin content of food, fruits and vegetables that go into storage with an especially high vitamin A or C content will come out of storage with a higher vitamin content than conventional varieties, providing they are stored properly.

The charts that follow list those varieties of vegetables and fruits that are generally recognized as being best for freezing, canning, pickling, drying, and keeping in some kind of cold storage, be it in a root cellar, basement, or outdoor area. High vitamin varieties are also listed. After each variety you'll find letters that represent seed companies in this country that sell that variety. By no means have we listed every seed company that sells these varieties. We have noted some of the larger companies located in different parts of the United States. We realize that many smaller seedsmen sell the same varieties and may also offer other varieties equally as good for these particular methods of storage. We also assume that many growers have had success preserving varieties different from those listed here. Although our listing is up-to-date at the time of publication, new varieties could possibly be added to the list now because improved varieties and hybrids are always being developed. We suggest that you contact your county agent and check the most current issues of seed catalogs for new varieties that are available to you.

The letters on the charts are abbreviations for the following seed companies:

Bg Burgess Seed & Plant Company
P. O. Box 218
Galesburg, MI 49053

Bp W. Atlee Burpee
Phila., PA 19132
Clinton, IA 52732
Riverside, CA 92502

E Earl May Seed & Nursery Company
Shenandoah, IA 51601

F Farmer Seed & Nursery
Faribault, MN 55021

G Gurney Seed & Nursery Company
Yankton, SD 57078

GB Gill Brothers Seed Company
Box 16128
Portland, OR 97216

H Joseph Harris Company, Inc.
Moreton Farm
Rochester, NY 14642

HF Henry Field Seed & Nursey Company
Shenandoah, IA 51601

J J.W. Jung Company
Randolph, WI 53956

M J.E. Miller
Canadaiqua, NY 14424

N The Natural Development Company
Bainbridge, PA 17502

P George Park Seed Company, Inc.
Greenwood, SC 29646

R Rayner Brothers
Salisbury, MD 21801

S R.H. Shumway
P. O. Box 777
Rockford, IL 61101

SB Stark Brothers Nurseries & Orchard Company
Rt. 2
Louisiana, MO 63353

SQ Schells Quality Seeds
Harrisburg, PA 17101

VEGETABLE VARIETIES

Variety	Use	Sources
Asparagus		
Mary Washington	good freezer	Bp, Bg, F, G, H, P, S, SB
California 500	good freezer	HF
Roberts	good freezer	E, G
Beans, bush, green		
Top Crop	good freezer and canner	Bg, Bp, E, F, G, HF, P, S
Royalty Purple	good freezer	Bg, F
Green Isle	good freezer and canner	F
Tenderette	good freezer and canner	G, HF, JH, S
Leka Lake	good freezer and canner	Bp
Improved Tendergreen	good freezer and canner	E, F, G, HF, S
Burpee's Stringless	good freezer	Bp, G, HF, P, S
Blue Lake	good freezer and canner	E, G, HF, J
Garden Green	good freezer and canner	G
(Early) Contender	good freezer and canner	E, G
Spartan Arrow	good freezer	F
Greencrop	good freezer	F
Wade	good freezer	F
Beans, bush, yellow		
Golden Wax	good freezer	G, HF
Pencil Pod Wax	good freezer	E, G, HF, J
Brittle Wax	good freezer and canner	Bp, FS, J
Kinghorn Wax	good freezer and canner	Bp, G, H
Cherokee Wax	good freezer	E, G, H
Butter Wax	good canner	E, G, H
Beans, pole		
Kentucky Wonder (Old Homestead)	good freezer	Bp, E, F, G, H, HF, J, P, S

Blue Lake	good freezer and canner	E, HF, J
Purple Pod	good canner	HF
New Pole	good freezer and canner	F
Beans, for drying		
White Mexican		S
Red Kidney		E, F, G, HF, S
Improved White Navy		E, G, S
Great Northern (White)		E, F, HF, S
White Marrowfat		S
Dwarf Horticultural or Wren's Egg		E, S
McClasan Pole		S
Black Eye		E, G, HF, S
White Wonder		HF
Michlite		F
Sanilac		J
Beans, lima		
Fordhook 242 (Fordhook Potato)	good freezer and canner	Bg, Bp, E, G, H, HF, J, S
Henderson's	good freezer and canner	Bp, E, HF, S
Improved Giant Bush	good freezer	Bp, E, S
Baby Bush	good freezer and canner	Bp, HF, J
Romano Italian	good freezer and canner	Bp, S
Burpee's Best	good freezer, high vitamin C content	Bp
Clark's Green Seeded	good freezer and canner	G
Beets		
Detroit Dark Red	good keeper	Bg, Bp, E, F, G, H, HF, J, S
Red Ball	good freezer and canner	Bp
Early Wonder or Model	good canner	E, F, G, HF, J, S
Dark Red Canner	good pickler	HF
Red Ruby Queen	good canner and pickler	E, G, H

(Hybrid) Pacemaker	good canner	G, H
Sweetheart	good freezer and canner, high vitamin content	F
Baby Canning	good pickler	E
Broccoli		
Green Sprouting or Calabrese	good freezer	Bg, F, G, HF
Zenith	good freezer	FM
Greenbud	good freezer	Bp
New Spartan	good freezer	E
Brussels Sprouts		
Jade Cross Hybrid	good freezer	Bp, G, H. P,
Cabbage		
Lightning Express	high vitamin content	Bg
Hybrid Stonehead	good kraut	Bg, F
Jumbo or Large Late Drumhead	good keeper (in underground storage)	Bg
Danish Ballhead or Roundhead	good keeper	Bg, Bp, E, G, HF, H, J
Copenhagen Market	good keeper	Bg, Bp, E, HF, S
Early Flat Dutch	good keeper	Bp, F, HF, S
Surehead	good keeper	Bp, HF
Premium Late Flat Dutch	good keeper	F, HF, S
Autumn King	good keeper	S
Wisconsin Red Hollander	good keeper	F, S
Wisconsin All Season	good keeper, kraut	F, S
Mammoth Red Rock	good keeper, kraut	E, F, G
Golden Acre	good kraut	E, F, G, H
Marion Market	good kraut	E, F
Penn State Ballhead	good keeper, kraut	F
Carrots		
Red Cored Chantenay	good freezer	Bp, F, G, HF, J
Danvers Half Long	good keeper (in underground storage)	Bp, E, G, F, HF, H, S

Nantes Half Long	good freezer and canner	Bp, H
Gold Pak	good keeper, freezer and canner	F, G, HF, J, S
Tendersweet	good freezer	E, G, HF
Royal Chantenay	good freezer	HF, H
Imperator	high vitamin C and carotene content	Bp, F
Morse Bunching	high vitamin C and carotene content	Bp, F
Coreless	good freezer	E, G, J
Cauliflower		
Early Snowball	good freezer and high vitamin A content	Bp, E, F, J, S
Super Snowball	good freezer	Bp, F, J
Purple Head	good freezer	F, H, S
Celery		
Fordhook	good keeper (in underground storage)	Bp
Giant Pascal	good keeper	Bg, Bp, S
Golden Self-Blanching	good keeper	Bg, F, HF, S, E
Corn		
Country Gentleman Hybrid	good canner	Bp, E, G, HF, S
Stowell's Evergreen	good freezer and canner	Bp, E, F, G, S
Golden Bantam	good freezer	Bp, E, G, HF, S
Golden Cross Bantam	good freezer and canner	Bp, F, G, HF, S
Illini or Xtra Sweet	good freezer	Bp, F, G, HF
Iochief	good canner	Bp, E, F, G, HF, H, S
Six Shooter Sugar Corn	good freezer	S
White Evergreen	good canner	S
Golden Delicious	good freezer	HF
Most Tender	good freezer and canner	HF

Early Golden Maincrop	good freezer and canner	HF
Hybrid Truckers	good freezer	G
Marcross Hybrid	good freezer and canner	G
Kanner King	good freezer and canner	F
Golden Beauty	good freezer and canner	F
Carmel Cross Hybrid	good freezer	G
Jubilee	good freezer and canner	F
Butter Nugget	good freezer	F
Early Sunglow Hybrid	good freezer	G
Wonderful	good freezer	H
New Cheddar Cross Hybrid	good freezer and canner	J
Kale		
Dwarf Blue Curled (Scotch)	high vitamin content	Bp, H, J
Onions		
Ringmaster	good keeper (in underground storage)	Bg, G
Hybrid Yellow Sweet Spanish	good keeper	Bg, Bp, E, F, G, HF, H, S
Crystal White Wax	good pickler	Bg, Bp
Ebenzer	good keeper	Bg, Bp, G, H, S
White Sweet Spanish	good keeper	G, S
Red Wethersfield Hamburger	good keeper	S
Southport Red, White and Yellow Globes	good keeper	E, F, G, H, J, S
White Sweet Slicer	good keeper	G
Downing Yellow Globe	good keeper	G, H
White Sweet Keeper	good keeper	E
Parsnips		
All America(n)	good keeper (in underground storage)	G, H

Peas		
Laxton's Progress	good freezer and canner	Bg, F, G, HF, S
Alaska	good canner and high vitamin C content	Bg, G, S
Green Arrow	good freezer	Bp
Little Marvel	good freezer and canner	Bg, Bp, E, F, G, H, HF, J, S
Giant Stride	good freezer and canner	Bg
Freezonian	good freezer and canner	Bg, Bp, F, HF, H, S
Wando	good freezer and canner	Bg, Bp, F, HF, H, J
Blue Bantam	good freezer	Bp, J
Tall Telephone	good freezer	F
Lincoln	good freezer	F, H
Burpeanna Early	good freezer	Bp
Sweet Green	good freezer	E, F
Early All-Sweet	good canner	G
Frosty	good freezer	E, G, H, P
Thomas Laxton	good freezer	Bp, G, S
Victory Freezer	good freezer and canner	G
Perfected Freezer	good freezer	H
Miragreen	good freezer	J
Progress No. 9 (Early Giant)	good freezer	E
Peppers, sweet		
Tasty	good freezer	F
Sweet Chocolate	good freezer	F
Worldbeater	good pickler and canner	Bp
Sunnybrook or Sweet Salad Tomato Pepper	good canner	S
Mammouth Ruby King	good pickler	S
Cherry Sweet	good pickler	HF
Early Thick Meat	good freezer	E
Peppers, hot		
Hungarian Wax or Yellow	good canner	Bg, Bp, G, HF, P, S

Variety	Use	Sources
Pimiento	good canner	Bg, E, HF, S
Red Chili	good pickler and drying	Bg, E
Long Hot Cayenne	good pickler	Bp, J, S
Anaheim Chilio	good pickler and drying	S
Jalapeno	good pickler	HF
Potatoes		
Norgold Russet	good keeper (in underground storage)	F, G, HF
Kennebec	good keeper, high vitamin C content	E, F, G, HF
Anoka	good freezer	F, G
Red Pontiac	good keeper	E, F, G, HF
White Cobbler	good keeper, high vitamin C content	E, G, HF
Pumpkins		
Small Sugar	good keeper (in underground storage)	Bg, Bp, HF
Jack O'Lantern	good canner	Bg, Bp, E, G, HF, J, S
Winter Luxury or Queen	good keeper	S
Yellow Connecticut Field or Big Tom	good canner	E, F, G, HF, H, J, S
Early Sweet Sugar	good canner	G
Radish		
Round White	good keeper (in underground storage)	Bg
White Chinese or Celestial Winter	good keeper	Bg, Bp
Comet	good keeper	Bp
Cherry Belle	good keeper	E, F, HF, H, J, S
White Giant or Hailstone	good keeper	HF, S, J
Stop Lite	good keeper	F
China Rose (Winter)	good keeper	E, HF, J, S
Round Black Spanish	good keeper	F, H, S
Long Black Spanish	good keeper	S

Rutabaga		
American Purple Top	good keeper (in underground storage)	E, G
Soybeans		
Bansei	good freezer	Bg, Bp, GB, N
Kanrich	high vitamin content	Bp, N
Spinach		
Bloomsdale (Long-Standing)	good freezer	Bp, E, F, HF, J, S
Hybrid No. 7	good freezer and canner	Bp, F
Virginia Blight Resistant	good freezer and canner	Bp
Giant Thick Leaf	good freezer and canner	HF
New Hybrid	good freezer and canner	G
America	good freezer	E
Northland	good freezer and canner	G
King of Denmark	good freezer	E
Sweet Potatoes		
Gold Rush	good keeper (in underground storage)	Bg
All Gold	high vitamin A content	F, E
Centennial	good keeper	E
Squash		
Table Queen, Acorn or Des Moines	good keeper (in underground storage)	Bg, Bp, F, G, HF, S, H, J, E
Hubbard (True or Warted)	good keeper	Bg, Bp, F, HF, H, J, E
(Golden or Red)	good keeper, freezer and canner	Bg, Bp, F
Sweet Meat Squash	good keeper	Bg
Banana Squash	good keeper	Bg, G, HF, S, J
Bush Scallop	good freezer	

Royal Acorn	good keeper	Bp
Buttercup	good keeper and freezer	Bg, Bp, F, G, S, H, J, E
Butternut	good keeper	Bg, Bp, F, G, H, J, E
Waltham Butternut	good keeper	Bp, F, HF, E
Prolific Straightneck	good freezer	H
Sweet Nut	high vitamin content	F
Gold Nugget	good keeper	F, G, H, J, E
Golden Delicious	good freezer and canner	G, H
Hybrid Gold	good keeper	G
Swiss Chard		
Lucullus	good freezer	Bg, Bp, F, G, S
Tomatoes		
Colossal	good canner	Bg
Pinkshipper	good canning	Bg
Rutgers	good canner and juice	Bg, Bp, F, HF, S, P, H, E
Marglobe	good canner	Bg, Bp, HF, S
Red (Yellow) Sugar	good canner, juice and preserves	Bg
Beefsteak	good canner	G, S
Abraham Lincoln	juice	S
Ponderosa	good canner	F, G, P, E
Roma (Red)	good canner, juice, sauce	Bp, S, H, E
Italian Canner	good canner, juice, sauce	Bp, S
Queen's	juice	S
Red Pear	good canner, preserves	HF, S
Yellow Pear	good canner, preserves	E
Garden State	good canner	S
Jubilee	juice	S
Tiny Tim Midget	preserves	F, J
Droplet	preserves	F
Bellarina	sauces, purees	G
Crimson Giant	good canner	G
Yellow Husk	preserves	G

Little Pear	preserves	G
Caro Red	high vitamin C content	Bg, J
Doublerich	high vitamin C content	Bg
Pink Gourmet	good canner, catsup	E
Turnips		
Purple Top White Globe	good keeper (in underground storage)	Bg, Bp, F, G, HF, H, J, E

FRUIT VARIETIES

Apples		
Anoka	good keeper (in underground storage)	G, HF, E
Fireside or Minnesota Delicious	good keeper	Bg, F
Northern Spy	good keeper, high vitamin C content	Bg, M, SB
Yellow (Red) Delicious	good keeper, sauce	Bg, FMG, HF, SB, E
McIntosh	good keeper, cider	F, G, HF, SB, J
Jonadel	good keeper, juice	F, G, HF, S, E
Transparent	sauce	E
Red Winesap (Crimson)	good keeper, juice, high vitamin C content, sauce	HF, R, SB, E
Chieftan	good keeper	HF, E
Stayman's Winesap	good keeper and freezer, sauce	HF
Jonathan	good keeper and freezer, sauce	G, HF
Grimes Gold	good keeper, sauce	HF
Haralson	good keeper	F, G
Baldwin	good keeper, high vitamin C content	M
Arkansas Black Twig	good keeper	SB
Prairie Spy	good keeper	F, G

Connell Red	good keeper	F, J
Rome Beauty	good keeper	SB
Red Dolgo Crab	pickler, jelly, sauce	Bg, F, G, S, E
Whitney Crab	pickler, jelly, sauce	F, G, HF, J, E
Blushing Golden	good keeper	SB
Splendor	good canner	SB
Jona delicious	good keeper	SB
Jon-a-Rich	good keeper	SB
Apricots		
Manchu	good canner, preserves	Bg, G, S
Hardy Iowa	good canner, preserves	HF
Hardy Superb	good canner, preserves	HF
Moongold	good canner, preserves	F, G, E
Scout	good canner, preserves	G
Blackberries		
Ebony King	good canner	G
Darrow	good canner, preserves	SB, E
Blueberries		
Mammouth Cultivated	preserves	S
Jersey	good freezer	
Coville	good freezer, preserves	HF, E
Saskatoon	preserves	G
Blue Ray	high vitamin C content	R, J, E
Rubel	high vitamin C content	R, Bg
Boysenberries		
New Thornless	good canner, preserves	Bg, G, E
Cherries, sour		
Early Richmond	good canner, preserves	Bg, HF, S, SB

Large Montmorency	good canner, preserves	Bg, G, HF, S
Meteor	good canner, preserves	Bg, F, HF, SB
Cherries, sweet		
Royal Ann or Napoleon	good freezer and canner	Bg, S, SB
Kansas Sweet	good freezer	Bg, HF
Yellow Glass	good freezer and canner	G
Bing	good freezer and canner	SB
Vista	good canner	SB
Cherries, bush		
Black Beauty	good canner	G, HF
Hansen	good canner, preserves	G, HF
Giant Red-Fleshed	good canner	HF
Nanking Cross	good canner, preserves	G
Drilea	preserves	G
Brooks	preserves	G
Currants		
Wilder Currant	jelly	G, HF
Red Lake	jelly	F, HF, E
Grapes		
Concord	juice, jelly	Bg, F, G, HF, S, SB, J
Seibel	juice, jelly	HF
Buffalo	juice, jelly	HF, SB
Red Caco	jelly	G, HF, J
Beta Grape	juice, jelly	F
Steubers	jelly	SB
Melons		
Crenshaw	good freezer	Bg, HF
Honey Dew	good keeper (in underground storage)	Bg
Hale's Best Muskmelon	good freezer	Bp, F, HF, J, E
Garrisonian Watermelon	good freezer	HF

Winter Melon	good keeper	F, G
Crimson Sweet	good keeper	E
Peaches		
Rochester	good keeper	Bg
Late Glo	good canner, freezer	SB
Eden	good canner	Bg, S
Golden Jubilee	good canner, pickler	Bg, HF, S
Wisconsin or Balmer	good canner, pickler	HF
Hale (Haven)	good canner	Bg, HF, SB, E
Ranger	good freezer and canner	SB
Fuzzless Gold	good keeper	HF
Polly	good canner	HF, E
Elberta (Queen)	good freezer and canner	Bg, G, SB, E
Pears		
Seckel	preserves	Bg, HF, SB, E
Duchess	preserves	S, SB
Colette	preserves	HF
Bartlett	good keeper (in underground storage), canner	Bg, G, HF, SB, J, E
Magness	good keeper and canner	HF, SB
Golden Spice	good pickler	G
Moorglow	good canner	SB
Tyson	good canner	SB
Persimmons		
Ozark	preserves	HF
Plums		
Blue Damson	good canner, preserves	Bg, HF, E
Mirabell	good canner	WG
Green Gage	good canner	S, SB
Superior	jelly, preserves	F, HF, E
Stanley Prune	good canner	Bg, F, HF, SB, E

Kaga	good canner, preserves	G
Idaho Prune	good canner	G
South Dakota	preserves	G
Sapalta	preserves	F
Mt. Royal Blue	preserves	Bg, F, J
Delicious	good canner, jelly	SB
Raspberries, red		
Indian Summer Everbearing	good canner	G, HF, S
Latham	good freezer and canner	Bg, F, S, SB, J
New Fall Red Everbearing	good freezer and canner	F, S, E
September Red Everbearing	jelly and preserves	Bg, G, SB, J
Raspberries, black		
Black Hawk	good freezer	Bg, G, HF, SB, J, E
Raspberries, purple		
Amethyst	good freezer	Bg, HF, S, E
Sodus	good freezer and canner, preserves	G, HF, J
Rhubarb		
Tenderstalk	good freezer	HF
Flare	good freezer	G
Victoria	good freezer	SB, E
Strawberries		
(Senator) Dunlap	good canner, preserves	F, G, HF, S, SB, J
Ozark Beauty Everbearing	good freezer and canner, preserves	Bg, F, G, HF, S, SB, J, E
Cyclone	good freezer	G, HF, E
Vesper	preserves	G
Surecrop	good freezer and canner	Bg, HF, SB, E
Streamliner Everbearing	good freezer, preserves	F, G, HF, E

Premier	good canner, preserves, high vitamin C content	F, R
Midway	good freezer, preserves	Bg, HF, J
Fairfax	high vitamin C content	R
Catskill	high vitamin C content	R
Midland	high vitamin C content	R
Tennessee Beauty	high vitamin C content	R
Robinson	high vitamin C content	R
Gem	high vitamin C content	R SQ
Blakemore	good freezer	HF
Ogallala Everbearing	good freezer, preserves, high vitamin C content	Bg, F, G, HF, E
Trumpeteer	good canner, freezer	F, G, E
Jumbo	good freezer, high vitamin C content	F
Sparkle	good freezer and canner, high vitamin C content	F, G, R, J, E
Wisconsin 537	good freezer	F, J
Paymaster	good freezer	F

VEGETABLES FOR VITAMINS

One of the important yardsticks for measuring the nutritional worth of any food is the contribution in terms of *vitamins* that it makes to our diet and our health. Vitamins are organic food substances—that is, substances existing only in living things, plant or animal. Although they exist in foods in minute quantities, they are absolutely necessary for proper growth and the maintenance of health. Plants manufacture their own vitamins. Animals obtain theirs from plants or from other animals that eat plants.

Vitamins are not foods in the sense that carbohydrates, fats, and proteins are foods. They are not needed in bulk to build muscle or tissue. However, they are essential, like hormones, in regulating body processes. As in the case of trace minerals (iodine, for instance), the presence or absence of vitamins in very small amounts means the difference between good and bad health. Many diseases and serious conditions in both human beings and animals are directly caused by a specific or combined vitamin deficiency.

The green leaves of plants are the laboratories in which plant vitamins are manufactured. So the green leaves and stalks of plants are full of vitamins. Foods that are seeds (beans, peas, kernels of wheat and corn, etc.) also contain vitamins which the plant has provided to nourish the next generation of plants. The lean meat of animals contains vitamins; the organs (heart, liver, etc.) contain even more, which the animal's digestive system has stored there. Milk and the yolk of eggs contain vitamins which the mother animal provides for her young. Fish store vitamins chiefly in their livers.

Basically there are two different kinds of vitamins—those that can be dissolved in fats and those that dissolve in water. The vitamins found in liver, eggs and butter are fat-soluble. Those in fruits and vegetables are watersoluble. Milk contains both kinds.

The vitamins we know most about are called by a letter and also a chemical name. These are vitamin A (carotene); vitamin

B_1 (thiamin), B_2 (riboflavin), B_6 (pyridoxine), the other members of the vitamin B group (biotin, choline, folic acid, inositol, niacin, pantothenic acid, para-aminobenzoic, B_{12}); vitamin C (ascorbic acid); the several D vitamins, D_2 (calciferol) and D_3 (7-dehydrocholesterol); vitamin E (tocopherol); vitamins F, K, L_1, L_2, M, P.

Researchers have established approximate estimates of the daily allowances of most of the vitamins for perfect health. These amounts are usually spoken in terms of milligrams. (A milligram is 1/1000 of a gram. A gram is 1/32 of an ounce.) You may also find daily vitamin allowances expressed in terms of International Units, which are each 1/6000 of a milligram.

Vegetables are an especially valuable source of many of the vitamins. Yellow and green leafy vegetables, along with tomatoes, contain appreciable amounts of carotene, the plant substance which is changed into vitamin A in the body. Ascorbic acid (vitamin C) is plentiful in tomatoes, peppers and many of the raw leafy vegetables. While potatoes have only a fair amount of ascorbic acid, the quantities in which they are eaten by many people make them a material source of it. Several of the B vitamins, too, are present in a variety of vegetables. The green leafy ones, legumes, peas, and potatoes provide some of the needed thiamin, riboflavin and niacin in a well-balanced diet.

Let's not forget that the *way* vegetables are grown has a definite role in the nutritive values—including vitamins—they will contain. As far back as 1939, the United States Department of Agriculture yearbook *Food and Life* stated:

> Underneath all agricultural practices there is a guiding principle . . . to carry out this cycle of destruction and construction economically—to see that plants, animals and man utilize raw materials efficiently to build up the products of life and that these products are broken down efficiently into raw materials that can be used again . . . By the proper use of fertilizers and other cultural practices it might be possible to insure the production of plant and animal products of better-than-average nutritive value for human beings.

VEGETABLES CONTAINING THE LARGEST AMOUNTS OF VITAMIN A

(Recommended Daily Allowance is 5,000 Units)

Vegetables	International Units of Vitamin A
Asparagus, fresh	1,000 in 12 stalks
Beans, snap	630 to 2,000 in 1 cup, cooked
Beet greens	6,700 in ½ cup, cooked
Broccoli	3,500 in 1 cup, cooked
Carrots, fresh	12,000 in 1 cup, cooked
Celery cabbage	9,000 in 1 cup
Collards	6,870 in 1 cup, cooked
Dandelion greens	13,650 in 1 cup, cooked
Endive (escarole)	10,000 to 15,000 in 1 head
Kale	7,540 in ½ cup, cooked
Lettuce, green	4,000 to 5,000 in 6 large leaves
Parsley	5,000 to 30,000 in 100 sprigs
Peas, split	1,680 in 1 pound
Peppers, green	3,000 in 2 peppers
Peppers, red	2,000 in 2 peppers
Pumpkin	1,200 to 3,400 in 1 cup, cooked
Spinach, fresh	9,420 in ½ cup, cooked
Spinach, canned	5,500 in ½ cup
Squash, winter	4,950 in ½ cup, cooked
Sweet potatoes	7,700 in 1 medium potato, baked
Tomatoes, fresh	1,100 in 1 medium tomato
Turnip greens	9,540 in ½ cup, cooked
Water Cress	4,000 in 1 bunch

In selecting vegetables, planning and varying those you serve at meals, it's important to remember that fresh vegetables offer higher overall nutrient values than either canned or frozen and in season are usually less expensive. Within economic reason, it is preferable that as much of any family's vegetable dietary as possible be of the fresh variety. Of course, the sooner they are eaten after picking or purchase, the more that vitamin and other food values are retained. Careful storing and refrigeration of those that must be held is essential for vitamin retention.

VEGETABLES CONTAINING LARGEST AMOUNTS OF B-COMPLEX VITAMINS

	Milligrams of			
Vegetable	Thiamin (B_1) (RDA is 5,000 units)	Riboflavin (B_2) (RDA is 1.2-1.8 mg)	Pyridoxine (B_6) (RDA is 2.0 mg)	Choline (RDA not established)
ARTICHOKES, Jerusalem				
ASPARAGUS, fresh	.16 in 12 stalks			130 in 12 stalks
BEANS, dried lima	.60 in ½ cup, cooked	.24 to .75 in ½ cup	55 in ½ cup	
BEANS, dried soy		.31 in ½ cup		340 in ½ cup
BEANS, green				340 in 1 cup
BEETS			.11 in ½ cup	8 in ½ cup
BEET TOPS		.17 to .30 in ½ cup		
BROCCOLI				
CABBAGE			.29 in ½ cup	250 in 1 cup
CARROTS				95 in 1 cup
CAULIFLOWER	.09 in ¼ small head			
COLLARDS	.19 in ½ cup, cooked	.22 in ½ cup, steamed		
CORN, yellow	1.9 in 1 lb.	0.5 in 1 lb.		
DANDELION, greens	.19 in 1 cup, steamed			
ENDIVE (escarole)		.20 in 1 head		
KALE				
MUSHROOMS				
MUSTARD GREENS		.20 to .37 in 1 cup, steamed		252 in 1 cup
PEAS, green, fresh	.36 in 1 cup, cooked	.18 to .21 in 1 cup, steamed	1.9 in 1 cup	260 in 1 cup
POTATOES, Irish			.16 in 1 medium potato	105 in 1 med. potato
POTATOES, sweet			.32 in 1 medium potato	35 in 1 med. potato
SPINACH		.24 to .30 in ½ cup, steamed		240 in ½ cup
TOMATOES, fresh				
TURNIPS			.10 in ½ cup	94 in ½ cup
TURNIP, greens	.10 in ½ cup, steamed	.35 to .56 in ½ cup		245 in ½ cup

Folic Acid (Micrograms) (RDA not established)	Inositol (RDA not established)	Niacin (RDA is 12-18 mg)	Pantothenic Acid (RDA not established)
			.40 in 4 artichokes
120 in 12 stalks			
330 in ½ cup	170 in ½ cup	9.6 in 1 lb.	.83 in ½ cup
71 in 1 cup			
42 in ½ cup, diced	21 in ½ cup		
25 in ½ cup			
90 to 110 in 1 cup			
	95 in 1 cup		1.4 in 1 cup
97 in 1 cup	48 in 1 cup		.2 in ½ cup
44 in ¼ small head	95 in ¼ head		.92 in ¼ head
		7.8 in 1 lb.	2.3 in 1 lb.
62 to 75 in 1 head			
100 in 1 cup			.30 in 1 cup
98 in ½ cup, cooked	17 in 7 mushrooms	6.0 in 7 mushrooms	1.7 in 7 mushrooms
22 in 1 cup	80 in ½ cup		.60 in 1 cup
140 in 1 small potato	29 in 1 med. potato		65 in 1 med. potato
	66 in 1 med. potato		.95 in 1 med. potato
225 to 280 in ½ cup	27 in ½ cup		
12 to 14 in 1 sm. tomato	46 in 1 sm. tomato		
	46 in ½ cup		

VEGETABLES CONTAINING THE LARGEST AMOUNTS OF VITAMIN C

(Recommended Daily Allowance is 70.75 mg.)

Vegetables	Milligrams of Vitamin C
Asparagus	20 in 8 stalks
Beans, green lima	42 in ½ cup
Beet greens, cooked	50 in ½ cup
Broccoli, flower	65 in ¾ cup
Broccoli, leaf	90 in ¾ cup
Brussels sprouts	130 in ¾ cup
Cabbage, raw	50 in 1 cup
Chard, Swiss, cooked	37 in ½ cup
Collards, cooked	70 in ½ cup
Dandelion greens, cooked	100 in 1 cup
Kale, cooked	96 in ¾ cup
Kohlrabi	50 in ½ cup
Leeks	25 in ½ cup
Mustard greens, cooked	125 in ½ cup
Parsley	70 in ½ cup
Parsnips	40 in ½ cup
Peas, fresh cooked	20 in 1 cup
Peppers, green	125 in 1 medium pepper
Peppers, pimento	100 in 1 medium pepper
Potatoes, sweet	25 in 1 medium potato
Potatoes, white, baked	20 in 1 medium potato
Potatoes, white, raw	33 in 1 medium potato
Radishes	25 in 15 large radishes
Rutabagas	26 in ¾ cup
Spinach, cooked	30 in ½ cup
Tomatoes, fresh	25 in 1 medium tomato
Turnips, cooked	22 in ½ cup
Turnips, raw	30 in 1 medium turnip
Turnip tops, cooked	130 in ½ cup
Watercress	54 in 1 average bunch

VEGETABLES CONTAINING THE LARGEST AMOUNTS OF VITAMIN E

(Recommended Daily Allowance has not been established)

Vegetables	Milligrams of Vitamin E
Beans, dry navy	3.60 in ½ cup steamed
Carrots	.45 in 1 cup
Celery	.48 in 1 cup
Lettuce	.50 in 6 large leaves
Onions	.26 in 2 medium raw onions
Peas, green	2.10 in 1 cup
Potatoes, white	.06 in 1 medium potato
Potatoes, sweet	4.0 in 1 medium potato
Tomatoes	.36 in 1 small tomato
Turnip greens	2.30 in ½ cup steamed

HARVESTING

If you grow your own food you've got it made over those who must rely on the grocery store or the supermarket for their daily sustenance, because you can pick and process the food that grows from your soil when its quality is at its very best. This means that you can harvest fruits and vegetables when they have reached just the right stage of maturity for eating, canning, freezing, drying, or underground storage, and you don't have to lose any time in getting the food from the ground into safe keeping, either.

The desired stage of maturity can vary from food to food and depends a great deal on what you intend to do with the produce once you've harvested it. In most cases, vegetables have their finest flavor when they are still young and tender: peas and corn while they taste sweet and not starchy; snap beans while the pods are tender and fleshy before the beans inside the pods get plump; summer squash while their skins are still soft. Carrots and beets have a sweeter flavor, and leafy vegetables will be crisp, but not tough and fibrous, when they are young. This is the stage at which you'll want to preserve their goodness.

Fruits, on the other hand, are usually at their best when ripe, for this is when their sugar and vitamin contents are at their peak. If you're going to can, freeze, dry, or store them, you'll want them firm and mature. But if you plan to use your fruits for jellies and preserves, you will not want them all fully ripe because their pectin content—that which helps them to gel—decreases as the fruit reaches maturity. In order to make better jellies some of the guavas, apples, plums, or currents you are using should be less than fully ripe.

It is nearly impossible to control just when your peaches, pears, apples and berries will be mature. Once planted, fruit trees and berry plants will bear their fruit year after year when the time is right. You're at their mercy and must be prepared to harvest just when the pickings are ready if you want to get the fruit at its best.

Vegetables are a different story. Because most are annuals and bear several weeks after they are planted, you can plan your garden to allow for succession plantings that extend the harvesting season for you and furnish you with a continued supply of fresh food at the right stage of maturity. This means eating fresh vegetables for the whole time your garden is producing and having vegetables just right for preserving several times—not all at once—for off-season months.

By planting three smaller crops of tomatoes instead of one large crop, you won't be deluged with more tomatoes than you can possibly eat and process at one time. Space your three pea plantings ten days apart in early spring and you'll have three harvests of peas and still plenty of time to plant a later crop of something else in the same plots after all the peas are picked. Vegetables, like salad greens, that do not keep well should be planted twice. Plant early lettuce about a month before the last frost and follow it with cauliflower. After the onions are out of the ground, put some fall lettuce in their place for September salads. If corn is one of your favorites and you've been waiting out the long winter for the first ears to come in, by all means, eat all the early-maturing corn you want, but make sure that enough late corn has been planted for freezing later on.

Vegetables that keep well stored fresh at low temperatures,

like cabbage, squash, and the root crops, should be harvested as late in the season as possible so you won't have to worry about keeping vegetables cool during a warm September or early October. Some vegetables, like carrots, parsnips, and Jerusalem artichokes, can be left right in the ground over the winter. It is wise to plant some late crops of these vegetables. Snap beans, planted in early May, can be followed by cabbage in mid-July. Beets planted in the beginning of April may be followed by carrots in July that can be stored right in the ground over the winter and into the early spring.

The charts that follow will give you a good idea as to the right time to harvest and how best to harvest for good eating and good keeping.

HARVESTING VEGETABLES

Vegetable	How and When to Harvest
Asparagus	Not until third year after planting when spears are 6 to 10 inches above ground while head is still tight. Harvest only 6 to 8 weeks to allow for sufficient top growth.
Beans, snap	Before pods are full size and while the seeds are about one-quarter developed, or about 2 to 3 weeks after first bloom. Snap pole beans just below the stem end, and you will be able to pick another bean from the same spot later in the season. Bush beans yield only one harvest, so it doesn't matter where you snap the pod from the bush. If you want to dry your beans, delay the harvest until the beans are dry on their stems.
Beans, lima	When the seeds are green and tender, just before they reach full size and plumpness. If you intend to dry your beans, harvest them when they are past the mature stage, when they are dry.
Beets	When 1¼ to 2 inches in diameter.
Broccoli	Before dark green blossom clusters begin to

open. Side heads will develop after central head is removed, until frost.

Brussels sprouts When full-sized and firm, before sprouts get yellow and tough. Lowest sprouts generally mature first. Sprouts may be picked for many months, even if temperatures go below freezing.

Cabbage When heads are solid and before they split. Splitting can be prevented by cutting or breaking off roots on one side with a spade after a rain.

Carrots When 1 to 1½ inches in diameter.

Cauliflower Before heads are ricey, discolored, or blemished. Tie outer leaves above the head when curds are 2 to 3 inches in diameter; heads will be ready in 4 to 12 days after tying.

Celery Self-blanching celery must be blanched 2 to 3 weeks before harvest in warm weather, and 3 weeks to a month in cool weather. If to be used immediately, cut plant root right below soil surface. For winter storage, plants are lifted with roots.

Chinese cabbage After heads form, cut as needed. For storage, pull up plants with roots attached.

Cucumbers When fruits are slender and dark green before color becomes lighter. Harvest daily at season's peak. If large cucumbers are allowed to develop and ripen, production will be reduced. For pickles, harvest when fruits have reached the desired size. Pick with a short piece of stem on each fruit.

Corn, sweet When kernels are fully filled out and in the milk stage as determined by thumbnail test. Use before kernels get doughy. Silks should be dry and brown, and tips of ears filled tight.

Eggplant	When fruits are half grown, before color becomes dull.
Endive (escarole)	To remove bitterness, blanch by tying outer leaves together when plants are 12 to 15 inches in diameter. Make sure plants are completely dry when this is done to prevent rot. Leave this way for 3 weeks before harvest. For storage, pull up plant with roots intact. Before a hard freeze.
Garlic	Pull when tops are dry and bent to the ground.
Kohlrabi	When balls are 2 to 3 inches in diameter. If bulbs are close together, cut them off below bulb in order not to cut off tangled roots of an adjacent bulb.
Jerusalem artichoke	Tubers can be dug anytime after the first frost and anytime throughout the winter.
Lettuce	Pick early in the day to preserve crispness caused by the cool night temperatures. Wash thoroughly, but briefly as soon as harvested, then dry to prevent vitamin loss. Loose leaf types, if cut off at ground level without disturbing roots, will send up new leaves for a second crop.
Okra	Pick a few days after flowers fall, while they are still young and not woody. Pods will be 1 to 4 inches long, depending upon variety. Freeze, can or dry at once because they quickly become woody once mature.
Onions	For storage, pull when tops fall over, shrivel at neck of the bulb and turn brown. Allow to mature fully but harvest before a heavy frost.
Parsnips	Delay harvest until a heavy frost. Roots may be safely left in ground over the winter and used the following spring before growth starts. (They are not poisonous if left in ground over winter.)

Peanuts In the South, dig vines before frost. Peanut shells should be veiny and show color, and the foilage slightly yellow. In the northern areas, peanuts are left in ground until mid-October.

Peas When pods are firm and well filled, but before the seeds reach their fullest size. For drying, allow them to dry on the bush.

Peppers When fruits are solid and have almost reached full size. For red peppers, allow fruits to become uniformly red.

Potatoes When tubers are large enough. Tubers continue to grow until vine dies. Skin on unripe tubers is thin and easily rubs off. For storage, potatoes should be mature and vines dead.

Pumpkins and squash Summer squash is harvested in early immature stage when skin is soft and before seeds ripen, before they are 8 inches long. Patty pans may be picked anytime from 1 to 4 inches in diameter. Skin should be soft enough to break easily with press of finger. If picked in this early stage, the vines will continue to bear. Winter squash and pumpkins should be well matured on the vine. Skin should be hard and not easily punctured with the thumbnail. Cut fruit off vine with a portion of the stem attached. Harvest before heavy frost.

Radishes Summer radishes should be pulled as soon as they reach a good size. Leaving them in the ground after maturity causes them to become bitter. Winter radishes may be left in the ground until after frost.

Rutabagas After exposure to frost but before heavy freeze.

Salsify Leave until after frost, as freezing of the roots improves flavor. Also may be dug out in spring.

Soybeans Green beans should be picked when the pods are almost mature, but before they start to yellow. Harvest period lasts for only about a week. Dry soybeans are allowed to dry on the vines and are picked just as they are dry, while stems are still green, otherwise the shells will shatter and drop their beans.

Spinach May be cut off entirely when fully mature (when 6 or more leaves are 7 inches long) or outer leaves can be cut from plant as they mature, letting inner leaves on to ripen.

Tomatoes When fruits are a uniform red, but before they become soft.

Turnips When 2 to 3 inches in diameter. Larger roots are coarse and bitter.

HARVESTING FRUITS

Fruit	How and When to Harvest
Apples	Summer apples are picked when ripe, and should be used or preserved at once. They usually do not store well for more than a few days. Pick fall and winter apples at peak ripeness for best storage. Pick with the stems; if stems are removed, a break in the skin is left which will let bacteria enter and cause rot. When picking apples, be careful not to break off the fruiting spur, which will bear fruit year after year if undamaged.
Apricot	Apricots should be left on the tree until fully ripe, because once they are picked they do not increase their supply of sugar. For drying, they should be ripe and firm.
Blackberries (and Boysenberries)	Berries are ripe when they fall readily from bush into the hand, a day or two after they blacken. Berries should be picked in the cool of the morning, kept out of the sun and refrigerated or processed as soon as possible.

Blueberries Blueberries should be left on the bush several days to a week after they turn blue. When fully ripe they are slightly soft, come easily from the bush and are sweet in flavor.

Cherries In order not to damage the fruiting twigs, cherries should be picked without the stems. This will leave a break in the fruit, therefore the picked cherries must be processed at once or else spoilage will begin. If sour cherries are protected from birds with netting, fruit should be allowed to ripen on tree for 2 to 3 weeks. The longer it hangs, the sweeter it becomes.

Currants The longer currants hang, the sweeter they become, so leave on bush for 4 to 6 weeks, unless you plan to make jelly from them. Then pick some of them when still a little green because they lose their pectin content as they ripen.

Dates In dry weather dates should be left on the trees until they are thoroughly ripe. If the weather becomes wet, they must be picked before any rain touches them. The ripening process is then finished indoors.

Gooseberries Because gooseberries of one variety all mature at the same time, they may be harvested in one day. The picker usually wears heavy leather gloves and strips the branches of their fruit by running his hand along the whole branch, catching the berries in an open container, such as a bushel basket. The small pieces of leaf and stem may be separated by rolling the fruit down a gentle incline made by tipping a piece of wood or cardboard. The leaves and stems will be left on the incline and the moderately clean fruit rolls to the bottom.

Grapes Grapes should be picked when fully ripe.

The fruit will be aromatic and sweet, and the stem of the bunch will begin to show brown areas. Grapes that are to be stored should be picked in the coolest part of the day. Clip the bunches from the stems with sharp shears and handle them by the stems rather than by the fruit.

Guavas Guavas ripen in a period of about 6 weeks. If they are to be used for jelly or juice, some of them may be picked before they are quite ripe. They contain such a large amount of pectin that a pound of fruit will make more than 3 pounds of jelly.

Oranges Color is not always a sign of maturity in oranges. Their skin contains a mixture of pigments: green, orange and yellow. In the fall the green predominates until the weather turns cool enough to check their growth, when the green fades out and the other pigments predominate. But in the spring when growth begins, green pigments will again appear in perfectly ripe fruit as it hangs on the tree. Navel oranges, Valencia and other varieties with tight skins are picked by pulling away from the stem. Those with loose skins, like mandarins, Temples and tangerines are picked with clippers which clip the fruit with about a half inch of the stem remaining. Once clipped from the tree, the remaining stub is clipped off.

Peaches Peaches are ripe when the fruit is yellow. Fully tree-ripened fruit will have more sugar and less acid than fruit which is picked when half ripe. Don't pull the fruit directly from the tree because it will cause a bruise which will make the fruit spoil quickly. Rather, remove the fruit from the tree by tipping and twisting it sideways.

Pears Pears are harvested when they have reached their full size and the skins change to a lighter green. Seeds will be starting to turn brown at this stage, and the stems separate easily from the tree when lifted. The quality of the pears will be much better if they are picked before stony granules are formed through the flesh, during the last few weeks of ripening.

Persimmons When fully ripe, persimmons are very soft and are very sweet. American persimmons may be harvested just before they are ripe or they may be left hanging on the tree into the winter months. Fruit left hanging through January, even though it is frozen on the tree, retains its flavor when picked and thawed.

Plums For canning and jelly-making plums may be harvested as soon as they have developed their bloom. At this point they are slightly soft, but still retain some of their tartness and firmness. Prune varieties, those that are to be dried, will hang on the tree long after they are ripe. They will develop more sugar as they hang.

Pomegranate Fruit is picked in the fall after it has changed color. It will ripen in cold storage. It may also be permitted to ripen on the tree, so long as it does not split.

Quince Quinces may hang on the bush until after the first fall frost. If they are to be stored, they may be picked a few weeks earlier.

Raspberries Because they become soft when fully ripe, raspberries must be picked every day, or at least every other day during harvesting time. Rain at harvest time causes berries to become moldy. They should therefore be picked immediately after a rain and processed at once before they mold. If moldy berries are left on the plant the mold will spread to green berries and destroy them.

Rosehips The fruit should be picked when it is fully mature in late fall. At this time the rosehips will be deep in color, have a mellow, nut-like taste, and the vitamin C content will be at its peak.

Strawberries The berries should be picked early in the morning when the fruit is still cool. Gently twist the fruit off its stem, do not pull it off the stem. Fruits washed without their stems will lose more vitamins than fruits which are de-stemmed after washing.

Watermelons Watermelons must ripen on the vine because they do not develop more sugar or better color after they have been taken from the vine while still green. Most melons are fully ripe when the tendril accompanying the fruits dies, but this is not always the case with all varieties. A ripe melon has a hollow sound, a green one, a metallic ring, when knocked with the knuckles.

HANDLING FOOD AFTER HARVEST

Making sure that you harvest your food at the right time is only half the key to great tasting fruits and vegetables kept through the winter months. Handling the food after the harvest—during the time between picking and processing—is just as important.

Although the actual growth of fruits and vegetables stops when they are plucked from the ground and cut off from their food supply, respiration and activity of enzymes continue. The physical and chemical qualities of the plants deteriorate rapidly. Not only will there be a deterioration of appearance and flavor as the freshness of food fades, there will also be a loss of nutrients particularly of vitamin C.

Fruits and vegetables should be prepared and canned, put into the freezer, dried, or placed in cold storage as soon as is humanly possible after harvest.

If you cannot avoid a delay in preparing and storing, cool your food as soon as it is picked. Do not keep it at room temperature, or, even worse, expose it to the sun. The quickest way to cool it is to immerse the food in ice water. After draining, keep the food at low temperatures, preferably between 32 and 40° F. Covering the produce with cracked ice is another means of cooling and thereby slowing down the loss of quality. These aids, of course, do not replace the need for prompt processing.

Fruits and vegetables are at their best when first picked. Don't expect any kind of storage to make a great food out of an inferior one. If you've taken the effort to grow good food, make the extra effort to harvest it at the right time and get it into proper storage as soon as possible.

FREEZING

For many, freezing is the best way of preserving the prides of their organic garden. Because you have to purchase a freezer and keep it running year round, freezing may cost you more money than other ways of preserving, but the money is well spent when you consider food flavor, color, texture, and nutrients, and the time you save by freezing rather than canning or drying your vegetables and fruits.

A general rule to remember is that those vegetables most suited for freezing are those which are usually cooked before serving. These include asparagus, lima beans, beets, beet greens, cauliflower, broccoli, Brussels sprouts, peas, carrots, kohlrabi, rhubarb, squash, sweet corn, spinach, and other vegetable greens. Vegetables that are usually eaten raw, such as celery, cabbage, cucumbers, lettuce, onions, radishes, and tomatoes, are least suited for freezing. Almost all fruits, especially berries, freeze very well.

If you are concerned about preserving the Vitamin C content in foods, do not can, but freeze those foods that supply us with most of our vitamin C. Fruits and vegetables rich in vitamin C include broccoli, spinach, cauliflower, Brussels sprouts, kohlrabi, turnip greens, strawberries, grapefruit, lemons, and oranges.

The freezing process itself does not destroy any nutrients in food. However, there can be some nutrient loss during blanching of vegetables and the cooling process that takes place right before food is frozen, when food combines with the oxygen of the air and goes through a process called oxidation, or from the "drip" which results from excessive thawing. Nutrient losses can be kept at a minimum if you are quick and efficient when blanching and freezing. Food should be prepared as soon as it is harvested, or kept at 40° F. or lower no longer than twenty-four hours before it is prepared and frozen.

Studies published by the North Dakota Cooperative Extension Service show that if foods are prepared and frozen properly, they retain their food value the same as fresh foods:

Carbohydrates show no change with the exception of the sugar (sucrose) being reduced to simple sugars (glucose and fructose) during long storage. This is of no importance.

Minerals might be lost in solution during the blanching and cooking of vegetables, but this loss is usually no greater than when cooking fresh vegetables.

Vitamin A is lost only when vegetables are not blanched.

B vitamins, such as thiamine and riboflavin, are lost by overstorage and through solution in cooling and blanching because they are water soluble. However, a greater loss is suffered if the vegetables are not blanched. Some thiamine is lost by the heat in cooking.

Vitamin C is easily lost through solution as well as oxidation. Vegetables lose some of their vitamin C through blanching, but without this process, the loss would be much greater. The same amount of vitamin C is lost in cooking fresh vegetables.

CONTAINERS TO USE FOR FREEZING

For the actual containers to be used in fruit and vegetable freezing, any may be used that will exclude air and prevent contamination and loss of moisture. Plastic containers are ideal. If tin cans with friction lids or glass jars are utilized, don't fill them to within more than an inch of the top. That space will allow room for the normal expansion of the food without accidents.

Fruits which are to be packaged dry and all vegetables can be placed in heavy plastic bags, heat-sealed with a hot iron or closed with a wire band at a point as close to the contents as possible so that there is a minimum of air in the package. Wax containers made especially for freezer use can also be used. It is a good idea to overwrap thin plastic packages with another plastic bag or stockinette to prevent the plastic material from tearing once inside the freezer. Oxygen entering frozen food through tears in packaging can ruin the best quality foods. To test for rips and holes, fill the bag with water. If you discover a leak, don't use the bag for freezer storage. Even the smallest hole can allow oxygen to enter and moisture to escape. Of course, plastic containers can be used over and over again. So can heavy-duty plastic bags, so long as they are in good shape.

Any container or bag that is strong, moisture-proof, and can be sealed to exclude air makes a good freezer container for freezing fruits and vegetables. Plastic containers, glass canning jars, and heavy-duty plastic bags are all popular freezer containers.

All freezer containers should be marked before they are put into the freezer. The type of food and the date it was frozen should be marked so that you can pick the kind of fruit or vegetable that you want at a glance and can use up those that are oldest first. If you are freezing different varieties of the same food, mark the variety on the label as well, so that you'll be able to determine for the next season which varieties freeze best.

THAWING FRUITS AND VEGETABLES

Freezing is not a method of sterilization as is heat processing in canning. While many microorganisms are destroyed by freezing temperatures, there are some, most notably molds, that continue to live even through their growth is retarded and their activity rate is slowed down. When foods are removed from freezer storage, their temperatures rise and the dormant microoganisms begin to multiply. Even during thawing the process of spoilage sets in, and the higher the thawing temperature, the faster the growth of spoilage microorganisms. Thus, the microbial population will increase at a slower rate if food is thawed at a low temperature, such as in a refrigerator, than at room temperature.

Frozen foods that need to be thawed, such as those that will be eaten raw or mixed with other foods for casseroles should be thawed in the refrigerator and not on the kitchen counter or in hot water, whenever possible. Because decomposition of thawed foods is more rapid than fresh foods, they should be used as quickly as possible after thawing. Of course, if frozen food is to be cooked, there is no reason to thaw it first. Just remove it from the freezer and immediately place it in boiling water or in a preheated oven and cook it frozen, before microorganisms become active.

HOW AND WHY TO BLANCH VEGETABLES

If you have ever tried to freeze vegetables without first blanching them, you may have discovered the horrible cardboard flavor they acquire after a few months in the freezer. They bear no relation to the succulence of the products you hopefully packed last summer. This is the work of enzymes.

Vegetables, as they come from the garden, have enzymes

working in them. These break down vitamin C in a short time and convert starch into sugar. They are all slowed down (not stopped) by cold temperatures, but they are destroyed by heat—by blanching.

The blanching idea isn't new. Methods for scalding or steaming fresh produce in preparation for freezing were introduced over forty years ago. Since then, however, food specialists have been discovering more about the unrealized and subtle effects of using this pre-cold-storage process. They have found, for example, that blanching makes certain enzymes inactive which would otherwise cause unnatural colors and disagreeable flavors and odors to develop while the foods remain frozen. Then, too, they've found that blanched vegetables are somewhat softened, so that they can be packed more easily and solidly into freezer containers.

The first and foremost reason for blanching, of course, is to help frozen produce keep better. What freezing—and the right preparation for it—does is literally to hold on to as much natural "freshness" as possible.

There is another benefit in the blanch-freeze method, a benefit that is tremendously important to those who actually eat the frozen vegetables. Experiments have shown that ascorbic acid—or, as it's more commonly known, vitamin C—is retained in much greater amounts in many of those blanched before freezing. Some held two, three, and even four times more of this elusive element through periods ranging up to nine months. Aside from maintaining better quality, this nutritive advantage over vegetables frozen unblanched (or treated in other ways, such as with sulfur dioxide gas) is significant.

Concentrated research on food freezing, blanching, and quality and nutrient retention was carried out at the University of Illinois College of Agriculture for several years. The University's experiment station quarterly, *Illinois Research* (Summer, 1959), reported extensive experiments on the blanching questions.

Five vegetables—broccoli, peas, snap beans, spinach, and corn—were picked fresh, at optimum maturity for freezing, and processed promptly. Several lots of each were used. Part of these were given preliminary preparation, blanching, cooling, packaging, freezing, and freezer storage, according to standard directions. The rest were packaged and frozen without blanching.

The green vegetables were compared for ascorbic acid retention after freezer storage periods of one, three, six, and nine months. In one instance (broccoli after one month) the amount was equal. In all the others, analyses showed *more vitamin C was retained in the blanched vegetables than in those frozen unblanched,* at every period of testing.

After three months, for example, blanched broccoli had held 64 percent of its raw ascorbic acid content, compared to 57 percent for the unblanched samples. At six months, the difference had widened to 60 to 40 percent; and by nine months, 54 to 36 percent.

Peas revealed even greater losses where blanching had been omitted. After a month, blanched specimens tested 70 percent vitamin C retention, against 63 percent for the unblanched. By the third month, it was 75 to 55; six months, 71 to 36; and nine months after freezing 70 percent to 37 percent.

With snap beans, a larger disparity was found right at the start. Blanched beans had held 85 percent of their initial ascorbic acid after a single month in the freezer, while those unblanched had dropped back to 58 percent. In three months it was 83 to 44; at six, 64 to 15; and at nine months the blanched beans still had 43 percent—and the unblanched just 3 percent.

Spinach was tested after two weeks' freezer storage and showed a 52 to 28 percent advantage for blanching. Corn, which is low in vitamin C to begin with, was not tested for retention, but was included in the cooking and palatability comparisons since it is one of the more popular vegetables for home freezing.

In addition to checking for ascorbic acid variation in the blanched and unblanched produce, the Illinois research sought to compare what is called *palatability,* too. This includes such factors as appearance, color, texture, flavor (or off-flavor), and general acceptability. "During the entire nine months," states the report, "scores of blanched samples were 4.0 to 4.4, corresponding to ratings of good to high good. After cooking, the blanched vegetables were bright green and tender, and had a good flavor. No off flavors were noted."

On the other hand, unblanched samples that had been stored for only one month received general acceptability scores of 2.4 to 2.9, which corresponds to ratings between poor and fair.

"Strong off-flavors developed and unblanched products lost color during the first month in the freezer."

Over the longer storage periods, the vegetables that were not blanched deteriorated still further, some of them so much that they were considered inedible. Most became faded, dull or gray; all became tough or fibrous; and some, broccoli especially, developed an objectionable hay-like flavor.

Tests with the frozen corn disclosed palatability ratings between good and very good for the blanched samples, while none of the unblanched was considered even fair. Although the appearance and color of the unblanched corn had held up, its flavor had become disagreeable and there was deterioration in texture.

Steps to Follow for Freezing Vegetables

What do these research results mean to you? Just this: to keep the largest possible amounts of vitamin C in the vegetables you freeze and to keep them tasty and appetizing, it pays to blanch carefully before freezing. Whether you want to store your own surplus for a healthful supply next winter or if you'd like to be sure of putting up enough naturally grown produce bought throughout this season, here are some tips on freezing:

1. Line up everything needed for blanching and freezing *first*. Nothing counts more than speed in holding on to freshness, taste, and nutritive value. Plan a family operation deep-freeze; have all hands on deck to help quickly, and arrange equipment and containers in advance for a smooth production.

2. Pick young tender vegetables for freezer storage; freezing doesn't improve poor-quality produce. As a rule, it is better to choose slightly immature produce over any that is fully ripe; avoid bruised, damaged, or overripe vegetables. Harvest in early morning. Try to include some of the tastiest early-season crops; don't wait only for later ones.

3. Blanch with care and without delay. Vegetables should be thoroughly cleaned, edible parts cut into pieces if desired, then heated to stop or slow down enzyme action. For scalding, use at least a gallon of water to each pound of vegetable, preheated to boiling point in a covered kettle or utensil (preferably stainless steel, glass, or earthenware). Steaming is better for some vegeta-

bles because it helps retain more nutritive value. Use a wire-mesh holder or cheesecloth bag over one inch of boiling water in an eight-quart pot. The same arrangement is handy for plunging vegetables into boiling water, one pound at a time. Start timing as soon as basket or bag is immersed or set in place for steaming. If you live in a high altitude area, add half a minute to blanching time for each 2,000 feet above sea level.

4. Cool quickly to stop cooking at the right point. Vegetables that are overblanched show in a loss of color, texture, flavor, and nutritive value. Plunge blanched vegetables into cold water (below 60° F.); ice water or cold running water will do best.

Organic gardener Ruth Tirrell cools her blanched vegetables by what she calls the waterless freezing method. Instead of placing vegetables under running water or soaking them, which can wash away water-soluble vitamins and minerals, she pours her vegetables into a flat pan which is resting on another pan filled with ice cubes and cold water. Once the vegetables are in this top pan, she covers them with an ice bag or plastic bag which is filled with ice cubes. Miss Tirrell cautions that this method of cooling blanched vegetables takes a lot of ice, so have a good quantity ready before you begin.

Whereas tin cans can be filled right to the rim, fill glass canning jars with food and boiling water only to within ½ to 1 inch of the top. This headspace allows for expansion of food and bubbling of liquid during processing.

5. Package at once in suitable containers. Glass jars require one to one and a half inches headspace; paper and plastic containers call for leaving a half-inch headspace, except for vegetables like asparagus and broccoli that pack loosely and need no extra room. Work out air pockets gently and seal tightly.

6. Label all frozen food packages; indicate vegetable, date of freezing, and variety. Serve in logical order—remember food value and appeal are gradually lowered by long storage. Maximum freezer periods for most vegetables is eight to twelve months. Except for spinach and corn on the cob, cook without thawing. Avoid overcooking.

As mentioned earlier in this chapter, almost every vegetable that can be cooked before eating may be frozen successfully. The list that follows tells how to prepare for freezer storage just about every vegetable found in an American garden. Although all the vegetables listed can be frozen, there will most likely be some here that you will prefer to store in some other manner. Potatoes, cabbage, root crops, and squash store well in a root cellar or an outdoor storage area. Beets are preferred canned by most people. Herbs are easy to dry and convenient to use in dehydrated form. If your freezer space is limited, you'd be wise to consider these

Blanch herbs before packing them for the freezer. Tie sprigs of fresh-picked herbs into small bunches and blanch them, one bunch at a time, in a steaming basket for 1 minute. Then plunge each bunch into ice water to cool it quickly. Shake off excess water, pack loosely in a plastic bag, and seal with wire twists.

alternate ways of storage. If you like to freeze mixed vegetables, we suggest that you cut and blanch each vegetable separately. Then mix and freeze them. Since blanching times vary according to the type of vegetable, blanching separately for the required time assures you that each vegetable is sufficiently precooked.

PREPARING VEGETABLES FOR FREEZING

Artichokes Select small artichokes or artichoke hearts. Cut off the top of the bud and trim down to a cone. Wash and blanch: 8 minutes in boiling water or 8 to 10 minutes in steam. Cool, pack, and freeze.

Asparagus Use young green stalks. Rinse and sort for size. Cut in convenient, equal lengths to fit container. Blanch in steam or water from 2 to 4 minutes, depending upon size of pieces. Cool, pack, and freeze.

Beans, lima Pick when pods are slightly rounded and bright green. Wash and blanch in steam or water for 4 minutes, drain and shell. Rinse shelled beans in cold water. No additional blanching is necessary. Pack, and freeze.

Beans, shelled Pick pods when they are well filled, bright green, and tender. Wash and blanch for 2 minutes in water or 2 or 3 minutes in steam. Cool, pack, and freeze.

Beans, snap Pick when pods are of desired length, but before seeds mature. Wash in cold water and drain. Snip ends and cut, if desired. Blanch in steam or water, 3 to 4 minutes depending upon size and maturity. Cool, pack, and freeze.

Beets Beets are usually preferred canned, but they may be frozen. Harvest while tender and mild flavored. Wash and leave ½ inch of the tops on. Cook whole until tender, skin and cut, if desired. No further blanching is necessary. Cool, pack, and freeze.

Broccoli Select well-formed heads. Buds that show yellow flowers are too mature and should not be frozen or canned. Rinse, peel, and trim. Split broccoli lengthwise into pieces not more than 1½ inches across. Blanch 5 minutes in steam or 3 minutes in boiling water. Cool, pack, and freeze.

Brussels sprouts Pick only green buds. Like broccoli, heads that are turning yellow are too mature to process. Rinse and trim. Remove outer leaves. Blanch in steam or water 3 to 5 minutes, depending upon size. Cool, pack, and freeze.

Cabbage Trim off outer leaves. You can shred for tight packing or cut into wedges. Blanch the shredded cabbage in boiling water for 1½ minutes, or in steam for 3 minutes. Wedges should be blanched in boiling water for 3 minutes or in steam for 4 minutes. Cool, pack, and freeze.

Carrots Root cellar storage is preferred to freezing, but they may be frozen. Harvest while still tender and mild flavored. Trim, wash, and peel. Small carrots may be frozen whole. Cut others into ¼ inch cubes or slices. Blanch in boiling water: 2 minutes for small pieces, 3 minutes for larger pieces, and 5 minutes for whole carrots; Blanch in steam: 4 minutes for small pieces and 5 minutes for larger ones. Cool, pack, and freeze.

Cauliflower Select well-formed heads free of blemishes. Wash and break into flowerets. Peel and split stems. Soak in salt water for 30 minutes. Blanch 4 minutes in steam or water. Cool, pack, and freeze.

Celery If your space is limited, this is one food to leave out of the freezer because it does not freeze very well. But, it may be frozen and then used in cooked casseroles, stews and soups. Select crisp stalks. Clean well and cut across the rib into 1 inch pieces. Blanch in boiling water 3 minutes or in steam for 4 minutes. Cool, pack, and freeze.

Eggplant Select firm, heavy fruit of uniform dark purple color. Harvest while seeds are tender. Wash, peel, and cut into ⅓ to ½ inch slices or cubes. Dip in solution of 1 tablespoon lemon juice to 1 quart water. Blanch 4 minutes in steam or water. Dip again in lemon juice solution after heating and cooling. Pack and freeze.

Herbs Harvest on a sunny morning right before plants blossom. Remove damaged portions and rinse under cold water. Blanch for 1 minute in steam. Cool, pack, and freeze.

Kohlrabi Harvest while tender and of mild flavor. Avoid any that are overmature. Wash and trim off trunk. Slice or dice in ½ inch pieces or smaller. Blanch in steam or water for 1 to 2 minutes, depending upon size of cubes or slices. Cool, pack, and freeze.

Mushrooms Select firm, tender mushrooms, small to medium size. Wash and cut off lower part of stems. Cut large mushrooms into pieces. Add ⅓ teaspoon lemon juice to 1 gallon of water. Blanch in boiling water: 2 minutes for small, whole mushrooms: 4 minutes for large, whole mushrooms, and 2 minutes for slices. Or blanch in steam: 3 minutes for small mushrooms and slices, and 5 minutes for large, whole mushrooms. Cool, pack, and freeze.

Okra Select young, tender pods. Wash and cut off stems so as not to rupture seed cells. Blanch 2 to 3 minutes in boiling water, or 5 minutes in steam. Cool; freeze whole or slice crosswise.

Parsnips Choose smooth roots. Woody roots should not be used for freezing; they will be tough and tasteless. Remove tops, wash and peel. Cut into slices or chunks. Blanch in boiling water or steam for 3 minutes. Cool, pack, and freeze.

Peas Pick when seeds become plump and pods are rounded. Freeze the same day they are harvested, as sugar is lost rapidly at room temperature. Discard immature and tough peas. Shell peas. Do not wash. Blanch 1½ minutes in steam or water. Cool, pack, and freeze.

Peppers, Sweet and Pimentos Select when fully ripe, either green or red varieties. Skin should be glossy and thick. Wash and halve. Remove seeds and pulp. Slice or dice. Peppers do not require blanching, but you may blanch for 2 minutes. This makes packing easier. Cool, pack, and freeze.

Pumpkin, Summer and Winter Squash Harvest when fully colored and when shell becomes hard on pumpkins and winter squash. Summer squash should be harvested before rind becomes hard. Wash, pare, and cut into small pieces. Cook winter squash and pumpkins completely before packing. Do not add seasoning. Blanch summer squash in steam or water 4 minutes, and blanch zucchinis 2 to 3 minutes, depending upon size. Slice summer squash ½ inch thick. Cool, pack, and freeze.

Soybeans Pick when pods are well rounded, but still green. Yellow pods are too mature for processing. Two to 3 days too long in the garden will result in overmaturity. Wash and blanch 5 minutes in steam or water before shelling. Cool and shell with pea sheller or by hand. Rinse shelled beans in cold water. No additional blanching is necessary. Pack and freeze.

Spinach and other greens (beet, dandelion turnip, carrot, collard, kale, etc.) Harvest while still small and tender. Cut before seed stalks appear. Harvest entire spinach plant. Use only tender center leaves from old kale and mustard plants. Select carrot, turnip, or beet leaves from young plants. Rinse well. Trim off leaves from center stalk. Trim off large midribs and leaf stems. Discard insect-eaten or injured leaves. Blanch 2 minutes in boiling water or 3 minutes in steam. Stir while blanching to prevent leaves from matting together. Cool, pack, and freeze.

Sweet Potatoes Use smooth, firm sweet potatoes. Wash and cook in water or bake at 350° F. until soft. Cool, and remove skins if you wish. Pack whole, sliced, or mashed. To each 3 cups pulp mix 2 tablespoons lemon or orange juice or dip whole or sliced potatoes in ½ cup lemon juice to 1 quart water to retain bright color. Cool, pack, and freeze.

Tomatoes Tomatoes can better than they freeze. If

you do freeze tomatoes, use them within a few months for best taste and nutritive value. Pack tomatoes whole in freezer containers without blanching, or stew the tomatoes before freezing. Stew tomatoes by cutting them into quarters and simmering them slowly in a heavy pot (without adding water) until soft, about 20 minutes. Stir continuously to avoid scorching. Seasonings may be added before tomatoes are chilled, packed, and frozen, but don't add bread crumbs to thicken the stewed tomatoes until they are heated before serving.

Turnips Use only young, tender roots. Cut off tops, wash and peel, Slice into lengthwise strips or chop into small cubes. Blanch in boiling water or steam 2½ minutes. Chill, pack, and freeze.

APPROXIMATE YIELD OF FROZEN VEGETABLES FROM FRESH VEGETABLES

Vegetable	Fresh, as Purchased or Picked	Frozen
Asparagus	1 crate (12 2-pound bunches)	15 to 22 pints
	1 to 1½ pounds	1 pint
Beans, lima (in pods)	1 bushel (32 pounds)	12 to 16 pints
	2 to 2½ pounds	1 pint
Beans, snap, green, and wax	1 bushel (30 pounds)	30 to 45 pints
	⅔ to 1 pound	1 pint
Beet greens	15 pounds	10 to 15 pints
	1 to 1½ pounds	1 pint
Beets (without tops)	1 bushel (52 pounds)	35 to 42 pints
	1¼ to 1½ pounds	1 pint
Broccoli	1 crate (25 pounds)	24 pints
	1 pound	1 pint
Brussels sprouts	4 quart boxes	6 pints
	1 pound	1 pint
Carrots (without tops)	1 bushel (50 pounds)	32 to 40 pints
	1¼ to 1½ pounds	1 pint

Vegetable	Fresh, as Purchased or Picked	Frozen
Cauliflower	2 medium heads 1⅓ pounds	3 pints 1 pint
Chard	1 bushel (12 pounds) 1 to 1½ pounds	8 to 12 pints 1 pint
Collards	1 bushel (12 pounds) 1 to 1½ pounds	8 to 12 pints 1 pint
Corn, sweet (in husks)	1 bushel (35 pounds) 2 to 2½ pounds	14 to 17 pints 1 pint
Eggplant	1 pound	1 pint
Kale	1 bushel (18 pounds) 1 to 1½ pounds	12 to 18 pints 1 pint
Mustard greens	1 bushel (12 pounds) 1 to 1½ pounds	8 to 12 pints 1 pint
Peas	1 bushel (30 pounds) 2 to 2½ pounds	12 to 15 pints 1 pint
Peppers, green	⅔ pound (3 peppers)	1 pint
Pumpkin	3 pounds	2 pints
Spinach	1 bushel (18 pounds) 1 to 1½ pounds	12 to 18 pints 1 pint
Squash, summer	1 bushel (40 pounds) 1 to 1¼ pounds	32 to 40 pints 1 pint
Squash, winter	3 pounds	2 pints
Sweet potatoes	⅔ pound	1 pint

U.S. Department of Agriculture

FREEZING FRUITS

Fruits lend themselves to freezing better, in most cases, than do vegetables, because fruits do not need to be blanched before they are frozen. It is true that certain changes in the texture of the frozen fruits, a cellular breakdown or softening, is similar to the changes in cooked fruits. Some fruits, such as papayas, pears, mangoes, bananas, watermelons, and avocadoes, are more subject to the loss of texture than others and cannot be frozen very satisfactorily. Whenever it is possible to freeze them however, fruits can be made to retain more nutritive value and flavor than by any other method of preservation, and the process usually takes about half as much labor as canning. Fruits can be frozen safely for up to one year.

Freezing Fruits with Honey

Fruits are usually frozen in one of two ways—dry or floated in a sweet syrup. Most information on freezing fruits recommends freezing them with dry sugar or mixing them in a syrup of water and sugar. Fruits held at freezing temperatures will keep without the aid of a sweetener, but they may lose some of their flavor, texture, and color when packed alone. If you want to add something to preserve the taste and appearance of your frozen fruit, don't use sugar, use honey. While sugar adds little or no nutritional value to foods in its refined or unrefined state, honey, if uncooked and unfiltered, contains important vitamins, minerals, and enzymes that do add food value to your frozen fruit.

In general, you can substitute honey for sugar in recipes for freezing fruit. Just cut down the amount of sweetener you use by one-half. This means that one-quarter to one-half cup of honey is mixed with one pint of dry fruit when the recipe calls for one-half to one cup of sugar for each pint of dry fruit.

If you want to freeze your fruits in a sweet syrup, make one with honey instead of sugar. A thin to medium honey syrup is best to use with most fruits. A thin syrup can be made by blending one cup of honey with three cups of very hot water. A medium syrup can be made by blending two cups of honey with two cups of very hot water. Chill all syrups before using them. Use enough syrup to completely cover the fruit. If the fruit is packed tightly enough, one-half cup chilled honey syrup should be sufficient for pint containers and one cup for quart containers.

Because honey has a flavor of its own while sugar does not, it is important to use mild-flavored, light honeys for freezing (and canning, for that matter) unless you enjoy the taste of light fruit-flavored honey instead of sweet frozen fruit. Early summer and spring honeys are generally milder than those collected in the fall. Clover, locust, and alfalfa honeys are certainly more suitable for freezing with fruit than dark honeys, like buckwheat.

Preventing Fruits from Darkening

Changes in color, flavor, aroma, and ascorbic acid (vitamin C) content of fruits during freezing and thawing are caused mainly by oxidation. No browning will occur in fruit tissues until practically all the ascorbic acid has been oxidized.

There are a few steps that you can take to preserve the qualities of fresh fruit. Freeze only mature fruit. Immature fruit is usually higher in tannins and other constituents involved in darkening; some contain compounds which become bitter during freezer storage and thawing. Also, handle fruits and fruit products quickly during preparation for freezing, packing, partial thawing, and serving to minimize exposure to air. Cut directly into syrup any fruit that is likely to discolor and place crumpled wax paper or foil on top to keep fruit under the syrup.

Pears, peaches, apricots, sweet cherries, and figs darken easily during freezing. To help retain their color, these fruits may be packed in a pectin pack. To make a pectin pack, boil for one minute one box of commercial pectin and one cup of water, stirring constantly. Add one-quarter cup honey, stirring until dissolved. Then add enough cold water to make two cups of syrup. Place fruit and syrup in a freezer container. All the fruit must be covered by the pectin syrup. To make certain that all the fruit is submerged, stuff a piece of crumpled wax paper or foil under the container lid. This will keep the fruit below the syrup.

The addition of natural ascorbic acid will also help to prevent discoloration. A tablespoon or two of rose hip concentrate, in liquid or powder form, may be added to pure honey or a honey syrup before it is poured over the packed fruit. Try these natural sources of ascorbic acid with a small quantity of fruit first. Freeze the fruit and, after a few days, thaw it and test it to determine the amount of ascorbic acid you should use to avoid browning. Ascorbic acid has a bitter taste, so use it sparingly. You may find that the amount of honey will have to be adjusted.

Thawing Frozen Fruits

In frozen fruit, the softening of the plant tissue permits a more rapid rate of spoilage after thawing than in the corresponding fresh fruit. Fruit should be thawed at a low temperature in its original container to best retain its nutritive value, appearance, and flavor. Fruit should be eaten before it is entirely thawed or immediately after it has thawed.

PREPARING FRUITS FOR FREEZING

Apples To freeze in slices, peel, core, and slice apples. Pack them dry or mix with 2 to 4 tablespoons of honey, mixed with 2 tablespoons of lemon juice or 1 teaspoon rose hip concentrate (to prevent browning). For sauce, pare, but leave skins on (if organically grown) and either grind whole in blender, or cook until soft in open kettle and put through food grinder. Add honey and lemon juice to taste, if desired.

Apricots Skins tend to toughen during freezing. Unless you plan to use the apricots for pies, peel them before freezing. Dip a few of them at a time into boiling water for 15 seconds or until skins loosen. Chill them quickly in ice water and peel. Cut in half and remove pits. Add a few pits to each container for flavor. Trickle honey thinned with warm water over fruit. Add 1 teaspoon rose hip concentrate if desired. Apricots may also be packed in a thin or medium honey syrup or in a pectin pack.

Avocadoes Choose those that are ripe and perfect. Peel, cut in halves, remove pits. Scoop out the pulp and mash it. Pack and freeze.

Blackberries If organically grown, pick out leaves and debris, but do not wash. Pack dry or trickle small amount of honey into container. Seal and shake until well mixed.

Blueberries If organically grown, do not wash. Pick out stems and leaves. If wild, scald for 1 minute to prevent toughening of skins. Pack dry or with small amount of honey trickled over fruit.

Cantaloupe Cut flesh in slices, cubes or balls. Add honey and lemon juice if desired. The texture of cantaloupe can best be captured if the fruit is served before entirely thawed.

Carambola Wash and slice. The tough rind is not softened by freezing, so it is best used as a garnish. Pack in thin honey syrup.

Cherries, bush Sort, wash, pit, or pack whole. Add honey to taste.

Cherries, sour Wash and chill in ice water before pitting to minimize loss of juice. Stem and pit. Add a small amount of pure honey. Mix and pack.

Cherries, sweet Wash and chill in ice water before pitting to minimize loss of juice. Add lemon juice or rose hip concentrate to hold color. Light varieties need more lemon juice or rose hip concentrate to retain color. Light varieties may also be packed in a pectin pack to retain their color.

Coconut Drain out milk. Cut away the hull, but not the smooth, inner skin which contains many minerals. Leave in large pieces or grate in blender or meat chopper. Pack dry. It will keep well for one year.

Cranberries Choose plump, glossy berries. Sort, wash and drain. Pack dry or make a puree by cooking berries in 1 cup water to each pint of berries until skins burst. Put in blender or through food mill and add honey syrup. Pack and freeze. For raw relish, grind 1 orange with 1 pound of berries. If they are organically grown, grate with skin on. If they have been sprayed, remove the skins and discard. One cup crushed pineapple may also be added.

Currants Choose the larger varieties for freezing. Stem and wash. Pack dry or add honey to taste.

Dates Choose ripe, firm fruit. Wash and remove pits. Pack whole or puree dates in blender or food mill.

Figs Wash, sort, cut off stems, peel, and leave whole or slice. Cover with a thin syrup. For crushed figs, wash and coarsely grind in blender. Add honey if desired. Figs may also be packed in a pectin pack.

Gooseberries Sort, remove stems and blossom ends. Pack dry or in a thin or medium honey syrup.

Grapefruit Peel and remove sections from heavy membrane. Smaller membranes may be left on; they contain important vitamins. Pack dry or add honey to taste.

Grapes Wash and stem. Leave seedless grapes whole. Cut in half and remove seeds from others. Pack dry or in thin syrup.

Guavas For puree, remove seedy portion and strain to remove seeds. Sweeten with honey, if desired. For slices, pare, halve and slice. Cover with a thin honey syrup.

Lychees Wash. Leave about ¼ inch stem on fruit. Pack dry.

Mulberries Wash, if necessary, and stem. Pack dry or in honey syrup.

Oranges Peel and remove sections from heavy membrane. Pack dry or add honey to taste.

Peaches Use only fruit that is ripe enough so that skins may be pulled off without blanching. To avoid discoloration, prepare only fruit enough for one container at a time. Wash, skin, pit, and freeze in halves or slices. Add honey mixed with a small amount of lemon juice or rose hip concentrate if desired. Peaches may also be packed in pectin pack.

Pears Pears retain better appearance and texture when they are canned. If you wish to freeze them, choose ripe, but firm (not hard) fruit.

To freeze strawberries in a honey syrup, slice them into a dish, and pour a light honey syrup over them. One-half cup of syrup should be sufficient for each quart of berries. Gently toss the berries so that all are coated with syrup. To protect the berries from being crushed in the freezer, pack them in rigid or semi-rigid containers.

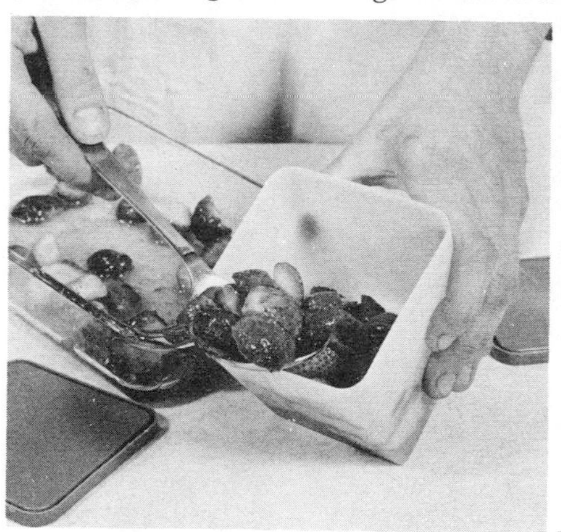

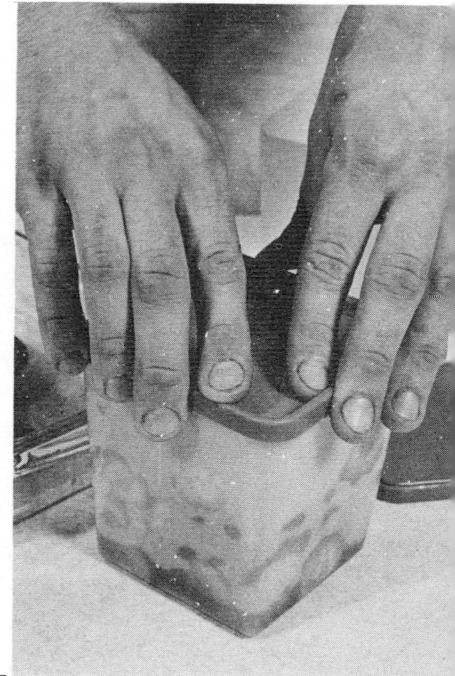

APPROXIMATE YIELD OF FROZEN FRUITS FROM FRESH

Fruit	Fresh, as Purchased or Picked	Frozen
Apples	1 bushel (48 pounds)	32 to 40 pints
	1 box (44 pounds)	29 to 35 pints
	1¼ to 1½ pounds	1 pint
Apricots	1 bushel (48 pounds)	60 to 72 pints
	1 crate (22 pounds)	28 to 33 pints
	2/3 to 4/5 pound	1 pint
Berries[1]	1 crate (24 quarts)	32 to 36 pints
	1⅓ to 1½ pints	1 pint
Cantaloupes	1 dozen (28 pounds)	22 pints
	1 to 1¼ pounds	1 pint
Cherries, sweet or sour	1 bushel (56 pounds)	36 to 44 pints
	1¼ to 1½ pounds	1 pint
Cranberries	1 box (25 pounds)	50 pints
	1 peck (8 pounds)	16 pints
	½ pound	1 pint
Currants	2 quarts (3 pounds)	4 pints
	¾ pound	1 pint
Peaches	1 bushel (48 pounds)	32 to 48 pints
	1 lug box (20 pounds)	13 to 20 pints
	1 to 1½ pounds	1 pint
Pears	1 bushel (50 pounds)	40 to 50 pints
	1 western box (46 pounds)	37 to 46 pints
	1 to 1¼ pounds	1 pint
Pineapple	5 pounds	4 pints
Plums and prunes	1 bushel (56 pounds)	38 to 56 pints
	1 crate (20 pounds)	13 to 20 pints
	1 to 1½ pounds	1 pint
Raspberries	1 crate (24 pounds)	24 pints
	1 pint	1 pint
Rhubarb	15 pounds	15 to 22 pints
	⅔ to 1 pound	1 pint
Strawberries	1 crate (24 quarts)	38 pints
	⅔ quart	1 pint

[1] Includes blackberries, blueberries, boysenberries, dewberries, elderberries, gooseberries, huckleberries, loganberries, and youngberries.

U.S. Department of Agriculture

Wash, peel and remove cores. Prepare only enough fruit at one time to fill a container to avoid unnecessary discoloration. Slice pears directly into a honey syrup mixed with a small amount of lemon juice or rose hip concentrate, or pack in a pectin pack.

Persimmons Sort, wash, slice, and freeze or press through a food mill for puree. Add 2 tablespoons of lemon juice per pint. Sweeten to taste with honey.

Pineapple Use only fruit ripened on the plant. Pare, trim, core, and slice or cut in wedges. Pack in own juice or in a thin honey syrup.

Plums and prunes If freestone, wash and pit, halve, or quarter. If clingstone, crush slightly, heat just to boiling, cool, and puree in food mill or blender. Add 2 tablespoons lemon juice or 1 teaspoon rose hip concentrate per pint. Sweeten with honey or add honey syrup.

Raspberries Clean and remove stems. Pack dry or fill containers and trickle in 2 tablespoons of honey per container. Seal and shake to mix.

Rhubarb Choose crisp tender red stalks. Early spring rhubarb freezes best. Remove leaves and discard any wooden ends. Wash and cut into 1-inch pieces. Pack dry or in a honey syrup.

Soursap Peel and cut lengthwise through the center. Remove and discard seeds. Force through food mill. Sweeten to taste with honey.

Strawberries Wash, slice, cut in half or freeze whole. Sweeten to taste with honey or use a thin honey syrup. One-half cup syrup is enough for one quart of berries. Strawberries may also be packed dry so long as you are freezing fully ripe berries.

WHAT TO DO WHEN THE FREEZER STOPS RUNNING

If, for any unfortunate reason (power failure, freezer breakdown, accidentally pulling the plug), your freezer stops functioning, don't run to it and open the door to check the contents. This is probably the worst thing you can do. First, try to approximate how long the freezer will be out of service. This may mean calling the electric company or the repairman. Do it as soon after you discover the problem as possible. If your freezer will be off for forty-eight hours or less and it is packed full, you have nothing to worry about. The food will stay frozen, so long as you don't open the door and let warm air enter. A freezer packed less than half full should cause more concern; it may not keep food frozen for more than twenty-four hours.

If you expect the food to start thawing before the freezer resumes operation, place dry ice (with gloves on—dry ice "burns") inside the freezer, out of direct contact with food. Place pieces of cardboard or wood on top of the food packages and put the dry ice on top of these. Close the freezer door and open only to add more ice. If the dry ice is placed in the freezer soon after it has stopped running, twenty-five pounds should keep the food frozen two to three days in a ten-cubic-foot freezer half full and three to four days in a fully-loaded freezer.

If it is necessary to remove some food to make room for the ice, store it in the refrigerator and cook the food as it thaws. Some foods moved to the refrigerator may be frozen again if your freezer starts operating sooner than you expected. Foods that do not reach refrigerator temperatures may be refrozen. If your fruits and vegetables are packed in plastic bags, squeeze them to see if they are still partially frozen. If they have not become soft and are even slightly firm and crunchy, you can safely refreeze them. Foods packed in rigid containers should be unsealed and inspected for ice crystals. If many remain, the food is cold enough to be refrozen. Mark refrozen foods appropriately and use them before you pull other foods from the freezer. Food that has been removed from the freezer and then cooked may be frozen again, so long as it is frozen soon after cooking.

If no dry ice is available, you can always move your frozen food to a commercial locker plant or to a friend's freezer. Pack the food in ice chests or wrap it in thick layers of newspaper to prevent thawing during transportation.

CANNING

Before freezers were around canning was the most popular method of preserving. In many households, especially in rural areas, canning is still the primary method of storing garden produce. More than 40 percent of all the families in the United States can food at home. An obvious advantage of canning is that there is practically no storage problem. You may can until your basement bulges, whereas your freezer space is definitely limited. And you need only invest in a pressure canner and canning jars (unless you plan to use tin cans and a sealer), both of which can be used over and over again, through many harvests.

STERILIZING FOOD

In simple terms, canning food means sterilizing it and keeping it sterile by sealing it in containers made from either glass or tin. Although commercial canning operations use tin cans almost exclusively, glass canning jars are preferred for home canning because they require no equipment for sealing. Sterilization is achieved by heating the food and the canning container sufficiently to kill all pathogenic and spoilage organisms that may be present in raw food. The duration of the processing time and the temperature at which the food and its container are held during processing depend upon the food being canned.

High-acid foods like pickled vegetables, fruits, and tomatoes are vulnerable only to heat-sensitive organisms. Boiling temperatures are sufficient enough to sterilize these foods and make them safe for consumption. Low-acid foods, like vegetables, meats, and dairy products, however, are not only susceptible to heat-sensitive organisms, but also to bacteria that can withstand temperatures above the boiling point of water. Clostridium botulinum forms a dangerous toxin that causes botulism, and this may be present in low-acid foods even after long boiling. Such foods must be processed at 240° F. to kill any possible traces of clostridium botulinum. Temperatures above the boiling point of water (212° F.) cannot be reached under ordinary conditions. A pressure canner must be used to process low-acid foods so that the necessary high temperature can be obtained.

Sealing the container immediately after heat processing makes it impossible for destructive organisms to invade the food and reinfect it. In addition, sealing creates a vacuum inside the container. This vacuum protects the color and flavor of the product, helps to retain the vitamin content, prevents rancidity due to oxidation, and assists in retarding corrosion of tin cans and corrosion of the closures on glass jars.

You can see why the success of preserving the taste, appearance, flavor, nutritional value, and safety of foods by canning depends upon the complete sterilization of foods and their containers and perfect seals. Extreme care must be taken to follow the canning instructions presented later in this chapter.

CHOOSING AND PREPARING YOUR FOOD

There are a few more things to remember when canning. As when freezing, use only fresh food in tip-top condition. Sort foods for size and maturity so that they will heat up evenly and pack well. Don't bring hot food into contact with copper, iron, or chipped enamelware. If possible, use soft water for syrups. Some foods darken or develop a gray tinge during the canning process. This is a chemical reaction between the food and the minerals in hard water or in metal utensils. Although such a discoloration does not mean that the food is unfit to eat, it may make canned food unattractive.

For the best results, food should be canned quickly, preferably on the day it is harvested. There should be no time lag between steps, so have all equipment clean and at hand before you begin. Food spoilage called "flat sour" can result if vegetables, particularly starchy ones, like corn, have stood too long between steps. The canned food may look all right and smell fine, but it has an unpleasant, sour taste. It is not fit to eat, even though it is not poisonous.

CHECKING SEALS

After processing and cooling canned foods, check their seals. Press down on the center of the lids on glass jars. If the lids do not "give" when you press on them, the jar is sealed properly. Check the tin cans by examining all seams and seals. Properly sealed cans should have flat, not bulging, ends, and the seams should be smooth with no buckling. If you suspect a container of having a faulty seal, don't take any chances. Either discard the food or open the container and process the food over again for the required time.

LABELING CONTAINERS

After processing and cooling the jars or cans, the type of food and the canning date should be marked on the top of the containers. Use the foods in the order in which they were canned. If you have canned different varieties of the same food, the variety of the food (for example: Peaches—Elberta, or Peaches—Hale) should be marked on the label so that you can compare different varieties and determine which varieties best retain their taste, flavor, and appearance after canning and storage.

STORING CANNED FOODS

Once canned, fruits and vegetables should be stored in a cool, dry place for best keeping. The higher the temperature of the storage area, the more chance of vitamin loss in the canned product. The U.S. Department of Agriculture tells us in the *Handbook of Agriculture* (1959) that canned fruits and vegetables will lose insignificant amounts of Vitamin C when stored at 65° F. Losses are about 2 to 7 percent after four months and increase slowly to about 10 percent after one year's time. When canned fruits and vegetables are stored at 80° F., however, 15 percent of the vitamin C value can be lost after four months, 20 percent after eight months and up to 25 percent after twelve months.

DON'T DISCARD THE LIQUID

The liquid in canned fruits and vegetables is an important source of food value. If you discard it, you're throwing out a good part of the vitamins and minerals found in the can or jar. Normally, fruit and vegetable solids make up about two-thirds of the total contents of the container; the rest is water. Soon after canning, the water soluble vitamins and minerals distribute themselves evenly throughout the solids and liquid. It follows, then, that about one-third of the water-soluble nutrients are in the liquid portion.

SPOILED FOOD

If, when opening canned food for use, you suspect that it has spoiled, do not test it by tasting it. Some spoilage bacteria, like those which cause botulism, are so toxic that a taste may be fatal. Boil the suspected food rapidly for a few minutes. If you notice an unusual and unappetizing odor developing, you can be certain that the food is not safe to eat. Burn the food or bury it deep enough so the no animal can uncover it and eat it.

CANNING VESSELS

High-acid foods, which include all fruits, tomatoes, and pickled vegetables, should be processed in a boiling-water canner. Any large vessel will do for a boiling-water canner so long as it meets these requirements: It should be deep enough to have at least one inch of water over the top of the jars and an inch or two extra space for boiling. It should have a snug-fitting cover. And there should be a rack to keep the jars from touching the bottom of the pot. If your steam-pressure canner is deep enough, you can use it for boiling-water processing. Set the cover in place without fastening it. Be sure to have the petcock open wide or remove the weighted gauge so that the steam escapes and no pressure is built up.

All vegetables, except tomatoes, are low-acid foods and must be processed in a steam-pressure canner because they require temperatures higher than that of boiling water for sterilization. Here you can't improvise; you must have a canner especially made for steam-pressure processing, and it must be in good working condition. The safety valve and petcock opening on the canner should be checked each time the steam-pressure canner is used to make sure that they are not stopped up with food or dirt. They can be cleaned by drawing a string or piece of cloth through them. Check the pressure gauge each year for accuracy. Instructions for checking should come with the canner. A weighted gauge needs only to be thoroughly cleaned. Make sure that the canner is clean before using it.

CONTAINERS

The following information about containers is excerpted from Circular 749 of the University of Illinois College of Agriculture:

Glass Jars. For processing foods in a boiling-water bath or a steam-pressure cooker, use only jars made especially for canning. Be sure all jars are in good condition, clean, and hot before packing food in them.

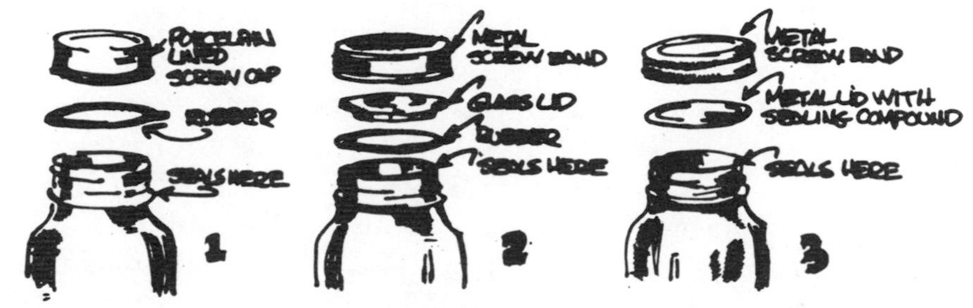

Three types of closures commonly used in canning foods: (1) mason top; (2) three-piece cap; and (3) two-piece cap.

There are three types of closures for glass jars. Be sure to follow the sealing directions that come with each type closure. Some general suggestions are given in the following paragraphs.

Mason Top. If porcelain lining is cracked, broken, or loose, or if there is even a slight dent at the seal edge, discard the cover. Opening these jars by thrusting a knife blade into the rubber and prying ruins many good covers. Each time you use a jar, have a new rubber ring of the right size.

Three-piece Cap fits a deep-thread jar with or without a shoulder. The metal band holds cap in place during processing and cooling. Remove it when contents of jar are cold, usually after twenty-four hours. Use a new rubber ring each time.

Two-piece Cap. Use the metal lid only once. The metal band is needed only during processing and cooling. Do not screw it farther after taking jar from canner. Remove band after contents of jar are cold, usually after twenty-four hours.

Tin Cans. Use only perfect cans and lids and gaskets. Wash cans in clean water and drain upside down. Do not wash lids, as washing may damage the gasket. If lids are dusty or dirty, wipe with a damp cloth just before putting them on the cans.

Tin cans need a sealer. Be sure the sealer you use is properly adjusted. To test the sealer, put a little water in a can, seal it, then submerge the can in boiling water for a few minutes. If air bubbles rise from the can, the seal is not tight and you will need to adjust the sealer, following manufacturer's directions.

RAW PACKS AND HOT PACKS FOR GLASS JARS

Both fruits and vegetables may be packed either by the raw-pack or hot-pack method. More food can be packed into the jars when the hot-pack is used because the food has already shrunk slightly during heating. Hot-packing is best used for foods that tend to discolor during canning. However, foods may lose some of their food value when they are hot-packed because of the additional heating. Both methods are described for individual fruits and vegetables in the charts that follow this section.

CANNING FRUITS

How to Prevent Fruits from Discoloring

Fruits have a tendency to darken in the canning process. To keep apricots, apples, peaches, and pears from darkening, they may be placed in a solution of two tablespoons lemon juice or vinegar and two tablespoons salt to a gallon of water. The fruits should be dropped in this solution as soon as they are washed and sliced. Let them soak for twenty minutes, then rinse them in cold water before packing.

Making Honey Syrup

Most people who can fruits pack them in a sweet syrup. However, fruit may be canned without sweetening. It may be packed in extracted juice or in water. Sugar helps canned fruit hold its shape, color, and flavor, but it is not needed to prevent spoilage. Unsweetened fruit should be processed just as sweetened fruit.

For your health's sake, if you wish to use a sweet syrup with your canned fruit, make it with honey, not with sugar or corn syrup. A very satisfactory syrup can be made by blending two cups honey with four cups of very hot water. Just as when freezing fruit with honey, choose a light-flavored honey for making your syrup so that the honey taste will not overpower the flavor of the fruit. To prevent the fruit from darkening, pack fruit in the hot honey syrup as soon as it is peeled and sliced.

Processing Fruits, Tomatoes, and Pickled Vegetables in Boiling-Water Bath

1. Fill boiling-water canner over half full, deep enough to cover containers. Turn on the heat.
2. If you are using jars, wash and rinse them and put them into hot water until needed. Pour boiling water over the lids and set them aside.

   If you are using tin cans and lids, have new ones ready, and have sealer at hand.
3. Prepare syrup, if it is to be used.
4. Prepare fruit for canning.
5. If you are using jars, pack fruit, either raw or hot, into jars and add hot syrup or water to fill jars no more than one-half inch from the top. Remove the air from the jars by running a spatula along the side, pressing fruit as you do so. Wipe the jars and screw the tops on tightly.

   If you are using tin cans, fill the cans with raw or hot fruit and cover the fruit with syrup or water. To exhaust the cans of air, the food inside must be heated to at least 170° F. Place the open, filled cans in a large pot with

TIMETABLE FOR—PROCESSING FRUITS, TOMATOES, AND PICKLED VEGETABLES IN BOILING-WATER BATH

PRODUCT	Raw pack or hot pack foods following directions. Put filled glass jars into canner containing hot or boiling water: For raw pack have water in canner hot but not boiling; for all other packs have water boiling. Add boiling water to bring water 1 or 2 inches over tops of jars but don't pour boiling water directly on glass jars. Put on cover of canner. Count processing time when water in canner comes to a rolling boil.	GLASS JARS		TIN CANS	
		PINTS	QUARTS	#2	#2½
		MIN.	MIN.	MIN.	MIN.
APPLES	HOT PACK 1. Pare, core, cut into pieces. To keep from darkening, place in water containing 2 tablespoons each of salt and vinegar per gallon. Drain, then boil 5 minutes in thin syrup or water. Pack apples in jars to ½ inch of top. Cover with hot syrup or water, leaving ½ inch at top.	15	20	10*	10*
	2. Make apple sauce, sweetened or unsweetened; pack hot to ¼ inch of top.	10	10	10*	10*
BEETS, PICKLED	HOT PACK Cut off beet tops, leaving 1 inch of stem and root. Wash beets, cover with boiling water and cook until tender. Remove skins and slice. For pickling syrup use 2 cups vinegar to 1 cup honey. Heat to boiling. Pack beets in jars to ½ inch of top. Add ½ teaspoon salt to pints, 1 teaspoon to quarts. Cover with boiling syrup, leaving ½ inch at top.	30	30		
BERRIES, EXCEPT STRAWBERRIES	RAW PACK Wash berries and drain. Fill jars to ½ inch of top, shaking berries down gently. Cover with boiling syrup (thin or medium recommended) leaving ½ inch at top.	10	15	15†	20†
	HOT PACK Wash berries and drain well. Add ¼ cup honey to each quart fruit. Cover pan and bring to boil. Pack berries to ½ inch of top.	10	15	15†	20†

* Use plain tin for apples, apricots, peaches, pears, sauerkraut, and tomatoes.
† Use R enamel cans for berries, cherries, plums, rhubarb.

CHERRIES	RAW PACK Wash; remove pits if desired. Fill jars to ½ inch of top, shaking cherries down gently. Cover with boiling syrup (thin or medium) leaving ½ inch at top.	20	25	20†	25†
	HOT PACK Wash; remove pits if desired. Add ¼ cup honey to each quart of fruit. Add a little water to unpitted cherries. Cover pan and bring to boil. Pack hot to ½ inch of top.	10	15	15†	20†
FRUIT PUREE	HOT PACK Use sound, ripe fruit. Wash; remove pits if desired. Cut large fruit in pieces. Simmer until soft, add a little water if needed. Put through strainer or food mill. Add honey to taste. Heat to simmering and pack to ¼ inch of top.	10	10	10†	10†
PEACHES OR APRICOTS	RAW PACK Wash peaches or apricots and remove skins. Remove pits. To keep from darkening place in solution (same as apples). Drain, pack fruit in jars to ½ inch of top. Cover with boiling syrup (light or medium) leaving ½ inch at top.	25	30	30*	35*
	HOT PACK Prepare fruit as for raw pack. Heat fruit through in hot syrup. If fruit is very juicy you may heat it with ¼ cup of honey to 1 quart of raw fruit adding no liquid. Pack fruit to ½ inch of top.				
PEARS	Peel, cut in halves, and core. Follow directions for peaches either raw pack or hot pack using same timetables.				
PLUMS	RAW PACK Wash. To can whole, prick skins. Freestone varieties may be halved and pitted. Pack fruit in jars to ½ inch of top. Cover with boiling syrup, leaving ½ inch space at top.	20	25	15†	20†
	HOT PACK Prepare as for raw pack. Heat to boiling in syrup or juice. If fruit is very juicy, you may heat it with honey, adding no liquid. Pack hot fruit to ½ inch of top. Cover with boiling syrup, leaving ½ inch at top.	20	25	15†	20†

TIMETABLE FOR—PROCESSING FRUITS, TOMATOES, AND PICKLED VEGETABLES IN BOILING-WATER BATH

PRODUCT	Raw pack or hot pack foods following directions. Put filled glass jars into canner containing hot or boiling water: For raw pack have water in canner hot but not boiling; for all other packs have water boiling. Add boiling water to bring water 1 or 2 inches over tops of jars but don't pour boiling water directly on glass jars. Put on cover of canner. Count processing time when water in canner comes to a rolling boil.	GLASS JARS		TIN CANS	
		PINTS	QUARTS	#2	#2½
RHUBARB	HOT PACK Wash and cut into ½ inch pieces. Add ¼ cup honey to each quart rhubarb and let stand to draw out juice. Bring to boiling. Pack hot to ½ inch of top.	MIN. 10	MIN. 10	MIN. 10†	MIN. 10†
SAUERKRAUT	HOT PACK Heat well-fermented sauerkraut to simmering (185°-210° F.). Pack hot kraut to ½ inch of top. Cover with hot juice, leaving ½ inch at top.	15	20	20*	25*
TOMATOES	RAW PACK Use only perfect, ripe tomatoes. Scald just long enough to loosen skins; plunge into cold water. Drain, peel, and core. Leave tomatoes whole or cut in halves or quarters. Pack tomatoes to ½ inch of top, pressing gently to fill spaces. Add ½ teaspoon salt to pints and 1 teaspoon to quarts, if desired.	40	50	55*	55*
	HOT PACK Quarter peeled tomatoes. Bring to boil and pack to ½ inch of top. Add salt as for raw packed tomatoes.	35	45	45*	45*

water about two inches below the tops of the cans. Cover the kettle and bring the water back to boiling. Boil until the food reaches 170° F. (about ten minutes). To be sure that the food is heated enough, test the temperature with a thermometer, placing the bulb in the center of one of the cans. Remove the cans from the water one at a time. Replace any liquid spilled from the cans by filling them with boiling syrup or water. Place a clean lid on each can and seal at once.

6. Place the closed jars or sealed cans upright in the canner. Water should be two inches over the tops of the containers. Add boiling water, if needed, being careful not to pour water directly on the containers.
7. Put the lid on the canner and bring the water to boiling.
8. Begin to count time as soon as the water starts to boil and process for time recommended in the charts that follows. (The processing times given on the chart are for altitudes less than 1,000 feet above sea level. For high altitude areas, increase the processing time one minute for each 1,000 feet above sea level. For example, if you live 3,000 feet above sea level, process three minutes longer than recommended time.) Boil gently and steadily, adding more boiling water as needed to keep containers covered.
9. As soon as processing time is up, remove the containers from the canner.
10. Unless the jars have self-sealing closures, complete the seals as soon as you take them out of the canner.
11. Set jars right side up to cool, making sure that they are far enough apart from one another so that air can circulate freely around them. For better circulation, place them on racks. Do not put them in a cold place or cover them while they are cooling.

    Tin cans should be cooled quickly in cold water, using as many changes of water as is necessary. Remove cans from the water while they are slightly warm so that they will dry in the air.

CANNING VEGETABLES

Preparing Your Vegetables

It is not necessary to pre-cook or blanch vegetables intended for canning since the enzymes that would otherwise break down the food are killed by the heat in the canning process.

Vegetables may be packed salted or unsalted. The small amount of salt used in canning does not prevent spoilage; it is only used as a seasoning. Can unsalted vegetables just as salted ones.

Processing Vegetables in the Steam-Pressure Canner

1. Put two or three inches of hot water in the bottom of the canner.
2. If you are using glass jars, wash and rinse them and put them into hot water until needed. Pour boiling water over the lids and set them aside.

   If you are using tin cans and lids, have new ones ready and have a sealer at hand.
3. Prepare vegetables for canning.
4. If you are using jars, pack vegetables, either raw or hot, into jars, leaving at least one-half-inch headspace. Pour in enough boiling water to cover vegetables. Wipe tops and threads of jars and screw tops on tightly.

   If you are using tin cans, fill the cans with raw or hot vegetables and cover them with water. To exhaust the air in the cans, the food must be heated to at least 170° F. Place the open, filled cans in a large pot with boiling water about two inches below the tops of the cans. Cover the pot and bring the water back to boiling. Boil until the food reaches 170° F. (about ten minutes). To be sure that the food is heated enough, test the temperature with a thermometer, placing the bulb in the center of one of the cans. Remove the cans from the water one at a time. Replace any liquid spilled from the cans by filling them with boiling water. Place a clean lid on each and seal at once.
5. Set the closed jars or sealed cans on a rack in the canner so that steam can circulate around them freely. Cans may

TIMETABLE FOR—PROCESSING LOW-ACID VEGETABLES

PRODUCT	Work rapidly. Raw pack or hot pack foods following directions, adding if desired ½ teaspoon salt for pints and 1 teaspoon for quarts. Place jars on rack in steam pressure canner containing 2 to 3 inches of boiling water. Fasten canner cover securely. Let steam escape 10 minutes or more before closing petcock.	USE 10 POUND PRESSURE				
		PRESSURE CANNER				PRESSURE SAUCEPAN
		GLASS JARS		TIN CANS		GLASS JARS
		PINTS	QUARTS	#2	#2½	PINTS
		MIN.	MIN.	MIN.	MIN.	MIN.
ASPARAGUS	RAW PACK Wash asparagus; trim off scales and tough ends and wash again. Cut in 1-inch pieces. Pack asparagus tightly as possible without crushing to ½ inch of top. Cover with boiling water leaving ½ inch at top.	25	30	20*	20*	45
	HOT PACK Prepare as for raw pack; then cover with boiling water. Boil 2 or 3 minutes. Pack asparagus loosely to ½ inch of top. Cover with boiling water leaving ½ inch at top.	25	30	20*	20*	45
BEANS, DRY WITH TOMATO OR MOLASSES SAUCE	HOT PACK Sort and wash dry beans. Cover with boiling water; boil 2 minutes, remove from heat and let soak 1 hour. Heat to boiling and drain, saving liquid for sauce. Fill jars ¾ full with hot beans. Add small piece of salt pork, ham, or bacon. Fill to ½ inch of top with hot tomato or molasses sauce.	65	75	65*	75*	85
BEANS, FRESH LIMA	RAW PACK Shell and wash beans. Pack loosely small type to 1 inch of top of jar for pints and 1½ inches for quarts; for large beans fill to ¾ inch of top for pints and 1¼ inches for quarts. Cover beans with boiling water.	40	50	40*	40*	60
	HOT PACK Shell the beans, then cover with boiling water, and bring to boil. Pack beans loosely in jar to 1 inch of top. Cover with boiling water, leaving 1 inch at top.	40	50	40*	40*	60

PRODUCT	Work rapidly. Raw pack or hot pack foods following directions, adding if desired ½ teaspoon salt for pints and 1 teaspoon for quarts. Place jars on rack in steam pressure canner containing 2 to 3 inches of boiling water. Fasten canner cover securely. Let steam escape 10 minutes or more before closing petcock.	USE 10 POUND PRESSURE				
		PRESSURE CANNER				PRESSURE SAUCEPAN
		GLASS JARS		TIN CANS		GLASS JARS
		PINTS	QUARTS	#2	#2½	PINTS
BEANS, SNAP	RAW PACK Wash beans. Trim ends and cut into 1 inch pieces. Pack tightly in jars to ½ inch of top. Cover with boiling water, leaving ½ inch at top.	MIN. 20	MIN. 25	MIN. 25*	MIN. 30*	MIN. 40
	HOT PACK Prepare as for raw pack beans. Then cover with boiling water and boil 5 minutes. Pack beans in jars loosely to ½ inch of top. Cover with boiling-hot cooking liquid and water, leaving ½ inch at top.	20	25	25*	30*	40
BEETS	HOT PACK Sort beets for size. Cut off tops, leaving 1 inch stem, also root; and wash. Boil until skins slip easily. Skin, trim, cut, and pack into jars to ½ inch of top. Cover with boiling water, leaving ½ inch at top.	30	35	30‡	30‡	50
CARROTS	RAW PACK Wash and scrape carrots. Slice, dice, or leave whole. Pack tightly in jars to 1 inch of top. Cover with boiling water.	25	30	25*	30*	45
	HOT PACK Prepare as for raw pack, then cover with boiling water and bring to boil. Pack carrots in jars to ½ inch of top. Cover with boiling-hot cooking liquid and water, leaving ½ inch at top.	25	30	20*	25*	45
CORN—CREAM STYLE	RAW PACK Husk corn and remove silk. Wash. Cut corn from cob at about center of kernel and scrape cobs. Pack corn loosely in pint jars to 1 inch of top. Cover with boiling water.	95		105†		115
	HOT PACK Prepare as for raw pack. Add 1 pint boiling water to each quart of corn. Heat to boiling. Pack hot corn to 1 inch of top.	85		105†		105
CORN—WHOLE KERNEL	RAW PACK Husk corn and remove silk. Wash. Cut from cob at about ⅔ the depth of	55	85§	60†	60†	75

CORN—WHOLE KERNEL, cont.	HOT PACK Prepare as for raw pack. To each quart of corn add 1 pint of boiling water. Heat to boiling. Pack loosely to 1 inch of top with mixture of corn and liquid.	55	85§	60†	60†	75
MUSHROOMS	HOT PACK Select tender, young mushrooms. Discard any that have opened. Wash thoroughly and trim off tough stalks. Cut in slices or leave small mushroom caps whole. Steam mushrooms for 4 minutes. Pack into hot jars, leaving 1 inch at top. You may add ½ teaspoon salt. Cover mushrooms with boiling water.					
PEAS, GREEN	RAW PACK Shell and wash peas. Pack peas loosely in jars to 1 inch of top. Cover with boiling water, leaving 1 inch at top.	40	40	30*	35*	60
	HOT PACK Prepare as for raw pack. Cover with boiling water and bring to boil. Pack peas loosely in jars to 1 inch of top. Cover with boiling water, leaving 1 inch at top.	40	40	30*	35*	60
PUMPKIN OR WINTER SQUASH CUBED	HOT PACK Wash pumpkin or winter squash, remove seeds, and pare. Cut into 1 inch cubes. Add just enough water to cover. Bring to boil. Pack cubes in jars to ½ inch of top. Cover with hot cooking liquid and water, leaving ½ inch at top.	55	90	50‡	75‡	75
PUMPKIN OR WINTER SQUASH STRAINED	HOT PACK Wash pumpkin or winter squash, remove seeds, and pare. Cut into 1 inch cubes. Steam until tender (about 25 minutes). Put through food mill or strainer. Simmer until heated. Pack hot in jars to ½ inch of top.	65	80	75‡	90‡	85
SPINACH AND OTHER GREENS	HOT PACK Pick over and wash thoroughly. Cut out tough stems and midribs. Place about 2½ pounds of spinach in cheesecloth bag and steam about 10 minutes or until well wilted. Pack loosely to ½ inch of top. Cover with boiling water, leaving ½ inch at top.	70	90	65*	75*	90

* Use plain tin.

† Use C enamel cans.
‡ Use R or sanitary enamel.

§ The State Department of Agriculture recommends all corn be canned in pints rather than quarts since processing time required for quarts tends to darken it. Courtesy North Carolina Extension Service.

be staggered without a rack between layers, but jars should have a metal rack between layers.

6. Fasten the cover of the canner securely so that no steam escapes except at the open petcock or weighted gauge opening.
7. Allow steam to escape from the opening for ten minutes so all the air is driven out of the canner. Then close the petcock or put on the weighted gauge and let the pressure rise to ten pounds. (If you live in a high-altitude area, you need to increase the pressure by one pound for each 2,000 feet above sea level. A weighted gauge may need to be corrected for high altitude by the manufacturer.)
8. Start counting time as soon as ten pounds pressure is reached, and process for the required time (see chart). Keep pressure as uniform as possible by regulating heat under canner.
9. At the end of processing time, gently remove canner from heat.
10. If you are using glass jars, let the canner stand until pressure returns to zero. Wait a minute or two, then slowly open petcock or remove weighted gauge.

Unfasten cover and tilt far side up so steam escapes away from you. As you take jars from canner, complete the seal if the jars are not the self-sealing type. Set jars upright on a rack, placing them far enough apart so that the air can circulate around all of them.

If you are using tin cans, release steam in canner at end of processing time by slowly opening petcock or taking off weighted gauge. When no more steam escapes, remove cover. Cool tin cans in cold water, changing water often enough to cool them quickly. Take cans out of cooling water while still slightly warm so they can air-dry.

PRESERVING VEGETABLES AND FRUITS BY DRYING

One of the oldest methods of preserving food has been largely abandoned in favor of modern food processing. But many people have rediscovered drying fruits, vegetables, and herbs as a perfect and natural storage method, as did the Indians, colonists, pioneers, and early farmers when our country was young.

Nothing is added, and only water is taken away. You can dry food from your garden with what was once common knowledge. Drying is not difficult, but to have a good dried product you must follow directions carefully. The faster you work, the higher will be the vitamin content of the dried food, and the better the flavor and cooking quality. In the drying process, most of the water content is, of course, removed. On a pound-for-pound basis, the dried food then has a substantially increased concentration of many nutrients, especially minerals. However, there is a loss of some vitamins, most notably, vitamins C and A. These two are usually significantly lowered in drying and subsequent storage, and this fact should be kept in mind when considering drying as a means of preservation.

Drying is especially popular with fruit. Apricots, peaches, pears, prunes, grapes, nectarines, figs, cherries, and apples are commonly dried. Dried vegetables are not as popular because most dried vegetables have limited uses. According to the Arkansas State Experimental Station, most vegetables, except for some varieties of beans and peas, show a significant loss in taste, color, and form when they are dried. Although most are not popular in dry form, most all vegetables can be dried. And all herbs can be preserved this way.

Both fruit and vegetables should be perfect for drying. Blemished or bruised fruit will not keep as well and may turn a whole tray of drying fruit bad. Fruit must be fully ripe so that its sugar content is at its peak. In most commercial drying operations, apricots and peaches are usually sulphured, but we disapprove of sulphuring fruit before it is dried. While sulphured fruit will retain more of its original color than unsulphured fruit and feel less leathery, sulphured fruit may impart a slight sour or acid taste and a questionable chemical to the diet.

Vegetables, except onions, which are used primarily for seasoning, and mushrooms, must be blanched or precooked in steam or boiling water after they are sliced for drying. This blanching sets the color, hastens drying by softening the tissues, checks the ripening process and prevents undesirable changes in flavor during drying and storage. Vegetables blanched before drying require less soaking before they are cooked for eating and have a better flavor and color when served. Blanching vegetables by steaming is preferable to blanching by boiling because nutrients are dissolved in the boiling water.

BLANCHING VEGETABLES

A pressure cooker or large, heavy pot makes a good steamer. Place a shallow layer of vegetables, not over two and one-half inches deep, in a wire basket or stainless steel or enamel colander. Have two or more inches of boiling water in the pot, set the basket on a rack above the water, cover tightly and keep the water boiling rapidly. Heat until every piece of vegetable is heated through.

If there is no convenient way of steaming, boiling is second best. Use a large amount of boiling water and a small amount of food so that the temperature of the water will not lower appreciably when the food is added. About three gallons of water to every quart of vegetables is good. Place the vegetables in a wire basket and immerse them in the boiling water for the required time, as suggested below.

PREPARING VEGETABLES FOR DRYING

Asparagus	Use only the top 3 inches of the spear. Blanch until tender and firm, about 10 minutes.
Beans, lima **Soybeans**	Shell and blanch 15 to 20 minutes.

Beans, snap Cut or slice into small pieces. Blanch about 20 minutes and place on trays about one-half inch deep.

Beets Remove tops and roots and blanch about 45 minutes or until cooked through. The time will depend upon the size of the beets. Cool, peel and cut into one-quarter-inch cubes or slice very thin.

Broccoli Trim and slice into small (one-half-inch) strips. Blanch 10 minutes.

Brussels sprouts Cut into lengthwise strips about one-half-inch thick; blanch 12 minutes and dry until crisp.

Cabbage Cut into long, thin slices and blanch 5 to 10 minutes.

Carrots Wash, slice or quarter. Blanch 8 to 12 minutes.

Celery Remove leaves and cut stalks into small pieces. Blanch about 10 minutes.

Corn Husk and remove the silk, then blanch 10 minutes to set the milk. Cut the kernels deep enough to obtain large grains, but be careful not to cut so deeply as to include any cob.

Mushrooms Peel and cut off stems if they are tough. Leave whole or slice, depending upon their size. Do not blanch, but dry while still raw.

Onions Peel and slice into small strips. Blanch 5 to 10 minutes, if they are to be eaten cooked. If for seasoning, do not blanch.

Peanuts Hang vines in well-ventilated shelter or attic, away from cold temperatures, or spread them on a drier, outdoors or inside. When dry and crunchy, they should cure for about 2 months before they can be eaten or roasted.

Peas Shell peas and blanch (15 minutes if steamed and 6 minutes if boiled).

Peppers Clean and slice into thin strips. Blanch 10 minutes.

Pumpkin, Winter squash Clean and cut into 1-inch strips and then peel. Blanch about 10 minutes, until slightly soft.

Rhubarb Cut into thin strips (about 1 inch wide) and blanch 3 minutes.

Spinach, Swiss chard, Kale Cut very coarsely into strips. Blanch spinach and Swiss chard about 5 minutes and Kale about 20 minutes. Spread not more than one-half-inch thick on trays

Summer squash, Zucchini Do not peel, slice into thin strips and blanch about 7 minutes.

Tomatoes Wash, quarter, and blanch for about 5 minutes. Run through a food mill to remove skins and seeds. Strain out the juice through a jelly bag or several layers of cheesecloth. Use a little hand pressure to extract more water, then spread the remaining pulp on glass, cookie sheets, or pieces of plastic. Turn the drying pulp frequently until it becomes dry flakes.

PREPARING FRUIT FOR DRYING

Apples Pare, core and cut into thin slices or rings.

Apricots Cut in half, remove pit and leave in halves or cut into slices or pieces.

Berries Halve strawberries and leave other, smaller berries whole.

Cherries Pit and remove stems. Let drain until no juice flows from them.

Grapes Remove stems and dip in boiling water to crack skins. Drain until no juice flows.

Peaches Cut in halves and remove pits. Skin and slice if desired.

Pears Skin, remove core. Cut into slices or rings.

Plums May be pitted or left whole. Blanch about 2 minutes to crack skin and make fruit dry quicker.

Prunes May be pitted or left whole. Like plums,

	they may be blanched two minutes to make drying easier.
Rosehips	Cut off blossom ends and stems.

DRYING TRAYS

Nearly all drying methods are quite simple. The sun and the air do almost all the work. Important safeguards are protection from insects and damp weather during the drying process. If you have a screened-in porch you're blessed.

Just about anything that has a good-sized flat surface can work as a drying tray, but the best trays are those that have ventilated bottoms made of wire mesh, cheesecloth, or wooden slats. If you're using solid boards, cover them with brown paper or clean grocery bags opened inside out. Trays made especially for drying food may be purchased, but they are easy to make. Construct wooden frames and stretch wire mesh or cheesecloth over them. The mesh or cheesecloth should be reinforced underneath with string that is tacked diagonally between the corners of each frame. If you're using cheesecloth, make the frames at least an inch or two under thirty-six inches, as most cheesecloth comes in bolts that are three feet wide. Put food on the tray, one piece deep and place a piece of cheesecloth or other fine-meshed ma-

You can make a drying tray such as this from spare lumber, string, and cheesecloth or wire mesh. Place food directly on the cheesecloth or wire mesh, and place the entire tray on rocks or wooden blocks to allow air to circulate over and under the food.

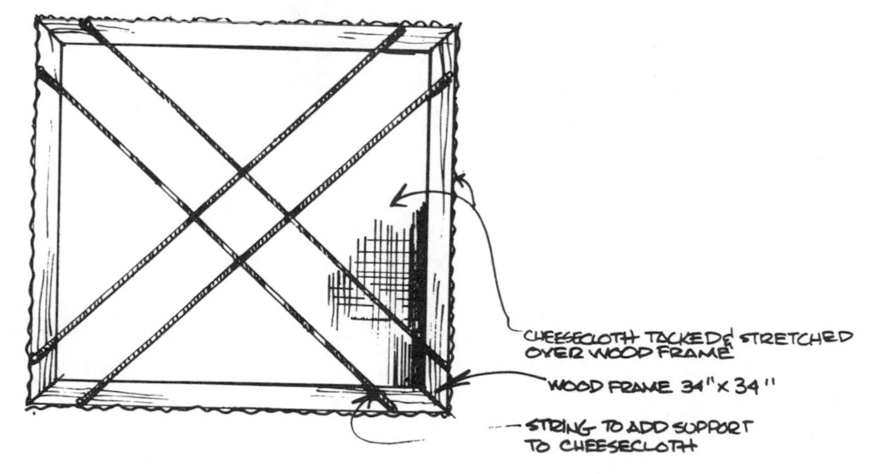

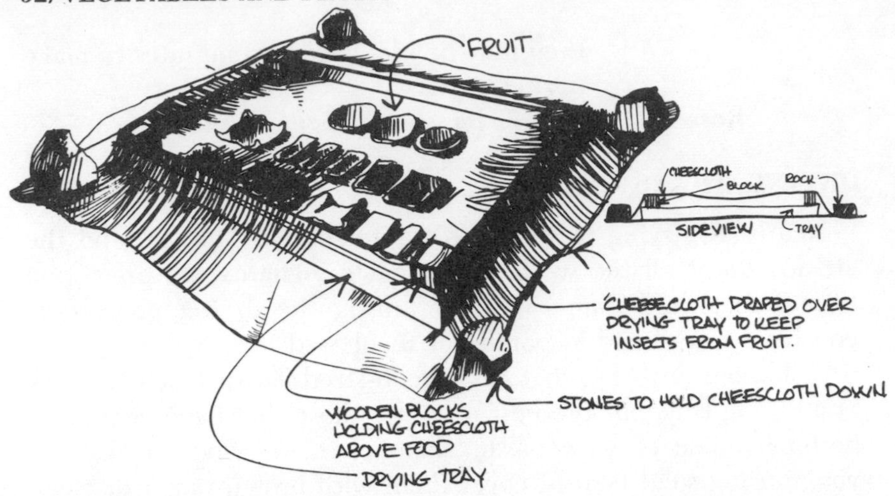

Such a drying set-up permits maximum air circulation while protecting fruit from insects. It does not, however, protect food from moisture. Put trays outside on sunny days after the morning dew has settled, and cover them or bring them indoors at dusk to protect the food from evening dew.

Clean window screens also make suitable drying trays. Place food only one piece deep on the screens or trays for best results.

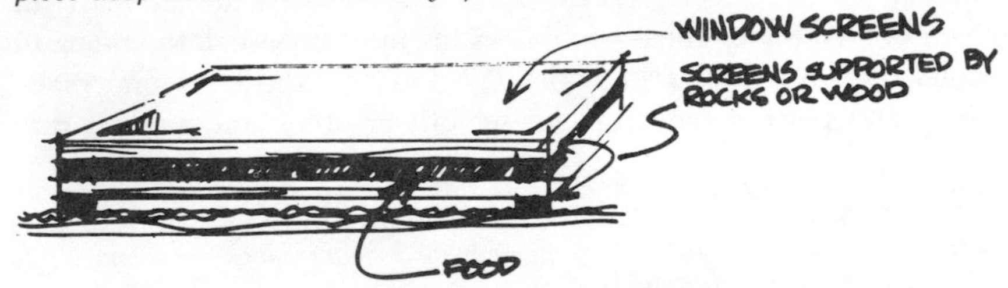

terial over—but not touching—the food, keeping it slightly above the food with blocks of wood or clean stones. When done out in the open, carefully lay strips of wood or stones on the cloth edges to prevent the material from blowing off. Pieces of fruit may also be dried between two clean window screens.

If drying is to be done outdoors, place the trays in a comparatively dust-free location, on racks raised above the ground. Raising the racks permits air to circulate freely under as well as over the food. If the trays are not protected by a sheet of thin (4-mil) polyethylene or glass, they must be covered at night to prevent dew from settling on the food. Be sure to exclude insects and animals from the drying area. Turn the food often, and make sure that you dry on sunny days.

DRYING INDOORS

When the atmosphere is dry and sunny, drying may be accomplished outside, but when heavy dews and frequent rains are normal, drying must be done indoors in an oven, in an attic, or in a specially constructed drier.

Drying with controlled heat in driers or in a kitchen oven has several advantages. The drying goes on day and night, in sunny or cloudy weather. Controlled heat driers shorten the drying time and extend the drying season to include late-maturing varieties. Vegetables dried with controlled heat cook up into more appetizing dishes than do sun-dried vegetables and have a higher vitamin A content and a better color and flavor.

Using the Oven

One oven can take about six pounds of prepared fruit or vegetable pieces at a time. Food should be exposed top and bottom. Place food directly on oven racks, one piece deep, or, if the slats are too far apart, cover them first with wire mesh and then place the food on top. Special drying trays, either purchased or made like those of wire mesh or wooden slats for drying food in the sun, may also be used in the oven. Separate trays in oven by placing three-inch blocks of wood at each corner when stacking them.

Constructing a Drier

There are many types of driers. Before building one, determine how many pounds of raw food it will hold and decide if it is large enough to be worthwhile. Each square foot of tray space will hold about one pound of raw prepared food.

California organic grower Elmer Kulsar made a simple and inexpensive construction that dried a five-acre crop of fruit with controlled heat from a kerosene heater. Mr. Kulsar began by building four rectangular frames made of 1-by-2 lumber that he nailed together with thin, cement-coated 6-penny nails. He tacked cheesecloth over the frames to hold the fruit. He then cut four legs from 2-by-2 stock, each 60 inches long, which allowed him to set the trays about 12 inches above the heater. He cut two pieces of ¼-inch plywood for the sides, each 24-by-34 inches,

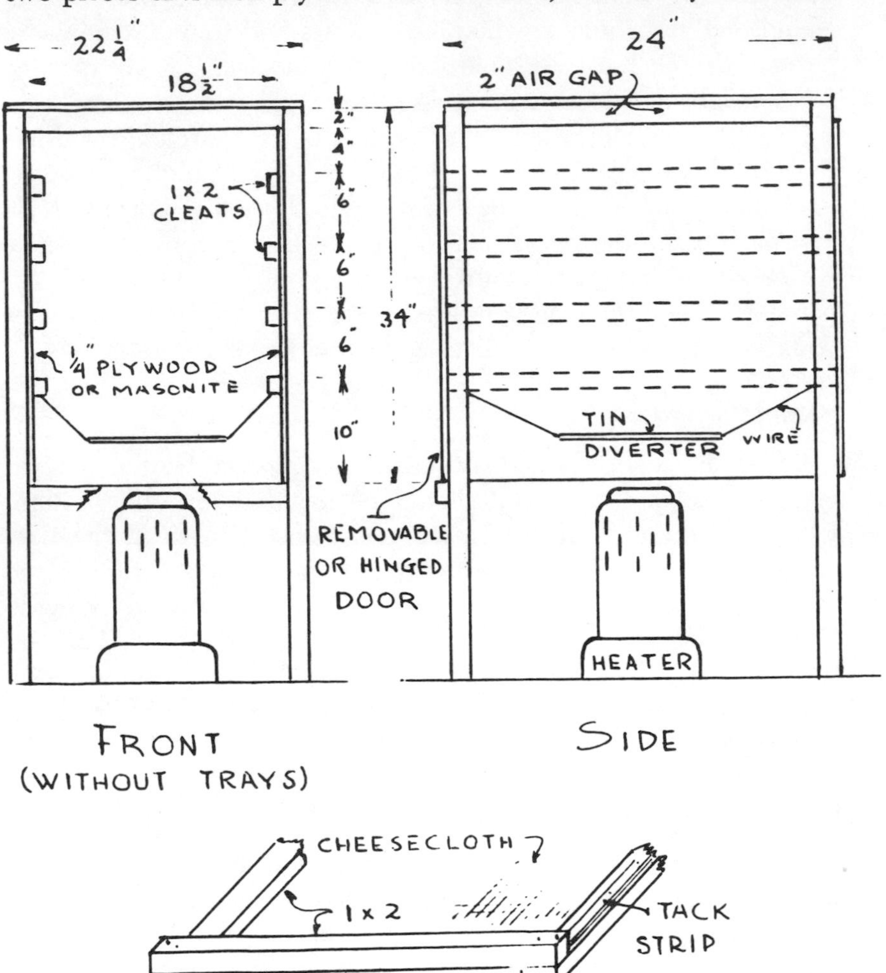

This 22-by-24-34-inch dryer holds 4 drying trays. The trays slide in and out easily so that those with the driest food can be shifted closest to the heater.

and nailed them flush to the legs, 24 inches apart and 2 inches down from the top, to allow the warm air to escape. He nailed eight 24-inch-long strips of 1-by-2 cleating to the sides 6 inches apart above the plywood to hold the trays securely. He made the 22-by-34-inch back out of ¼-inch plywood, and nailed it in place, again leaving a 2-inch gap at the top and spacing the sides 18½ inches apart so the tray could slide easily. The top, cut 22-by-24 inches, also allowed an 18½-inch space between the sides.

To complete the hastily-made, but effective, dryer, he added a 22-inch spacer strip to secure the front legs, and hung a 22-by-34-inch front door on a pair of nails. He then set the heater in position and hung sheet metal to serve as a heat diverter.

Any type of heater can be used, providing it supplies a steady flow of heat at a moderate and fairly even temperature. An old gas plate, electric hot plate, electric heat lamp, or kerosene heater (clean and adjusted so as not to smoke) will dry food satisfactorily. Too much heat is bad, so the distance under the bottom tray will have to be adjusted. Drying time can be cut down, and production stepped up, if the shrinking food is periodically moved closer together on the trays and shifted downward from the upper trays. Prepared food is then added to the top tray.

WHEN YOUR FOOD IS DRY

Drying is finished when fruit feels dry and leathery on the outside, but slightly moist inside; vegetables should be brittle. If in doubt, leave the food on the trays a little longer, but reduce the temperature if you're drying with the oven or a drier. Fruit seems to be moister when it is hot, so remove a few pieces from the tray occasionally and allow them to cool before you determine if they are dry. Most vegetables take four to twelve hours to dry. Fruits take six hours or longer, depending upon the size of the pieces and the drying conditions. Since some pieces of food will dry faster than others, it is important to remove pieces as they dry rather than wait until every piece of fruit or vegetable is totally dehydrated to stop the drying process. Food that overheats near the end of drying will scorch easily.

After the drying is finished, store the fruit in glass or airtight freezer bags that are closed tightly with wire or rubber bands.

For four successive days stir the contents thoroughly each day to bring the drier particles in contact with some that are more moist. In this way the moisture content will be evenly distributed. If, at the end of the four days, the fruit seems too moist, return it to the drier for further treatment. When well dried and conditioned, the fruit should be stored in a cool, dry basement or pantry. During warm, humid weather, dried foods retain their quality best if they are kept under refrigeration. It is best to examine dried food occasionally for mold. The danger of mold is prevented if the dried product can be stored at freezing temperatures or below.

If you discover bugs or worms in your fruit in late winter or early spring don't throw it out. Spread it in shallow pans, and put it in your oven for about twenty to twenty-five minutes at 300° F. The heat will take care of the vermin and sterilize your fruit at the same time. You can also heat fruit in a low oven if it becomes too limber and moist during storage. If all insects or insect eggs have been excluded, the food will remain good for many months, at least until the next harvest.

REHYDRATING

Water is taken out of fruits and vegetables for preservation, and, in most cases, you'll want to put water back in before you eat the food. Dried fruits are quite good eaten just as they are—by themselves or chopped up in cereals and desserts—but vegetables and fruits intended for baked products and compotes should be rehydrated or given back the water lost during drying.

To rehydrate fruits and vegetables pour one and one-half cups of boiling water over each cup of dried food. Let the mixture stand until all the water is absorbed. Vegetables generally absorb all the water they are capable of retaining in about two hours. Fruits require a longer soaking time—anywhere from two to several hours. Overnight soaking may be necessary for complete rehydration of some fruits. The amount of water dried foods will absorb and the time it takes for complete rehydration varies according to the size of the food and its degree of dryness. If the water is absorbed quickly, add more—a little at a time—until the food will hold no more.

Rehydrated fruits need not be cooked (unless, of course, you prefer them that way, but vegetables are always cooked after they have been soaked. To cook the vegetables, put them and any water they did not absorb in a pot. Add only enough extra water to cover the bottom of the pot. Cover and quickly bring the vegetables to a boil. Reduce the heat and simmer until the vegetables are plump and tender. If the vegetables are still tough after about five minutes of cooking or all the water is quickly absorbed by the cooking vegetables, they have not been soaked long enough. Next time, extend the soaking time so that the vegetables are fully rehydrated before cooking. Fruits are cooked in the same manner.

FRUIT LEATHER

The pulp from juicing peaches, apricots, prunes, and apples can be dried to form a naturally sweet, confection-like fruit leather that will keep in good condition for one year or more.

Apricot or Peach Leather

- 1 gallon pitted apricots or peaches
- 1½ cups unsweetened pineapple juice
- 3 teaspoons almond extract
- honey

Place the pitted fruit and pineapple juice in a large, heavy pot. Cover the pot and set it over low heat. Steam the fruit until it is soft. Drain off the juice well, lifting the fruit from the sides of the strainer to allow all the juice to run out freely. The more juice strained out, the quicker the process of "leather-making."

Run the fruit through a food mill or sieve, removing the skins if you prefer a smooth product, or use the skins as part of the pulp for the leather. Sweeten the pulp to taste with honey and the almond extract. The pulp should be as thick as apple butter or more so. Spread it on lightly oiled cookie sheets so that it is one-quarter of an inch thick. If it is much thicker than this, it will take very long to dry. Cover the cookie sheets with a single layer of cheesecloth to keep out dust and insects, and place them in a warm dry place to dry. Depending upon the weather, the pulp will dry in one to two weeks. Drying can be hastened by placing the cookie sheets in a low oven. Turn the oven con-

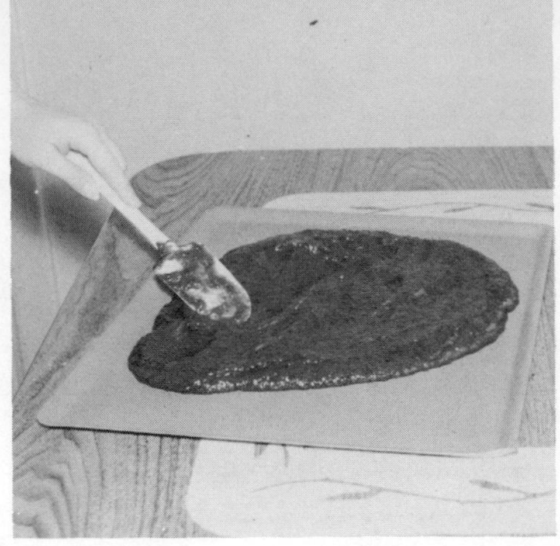

To make fruit leather, spread the sweetened fruit pulp ¼-inch thick on a lightly oiled cookie sheet. After about 2 weeks at room temperature, the pulp should be dry enough to be pulled from the cookie sheet in one solid, thin layer.

trol to warm (120° F.) and leave the oven door slightly open to allow moisture to escape.

When the leather is dry enough to be lifted or gently pulled from the cookie sheets, put the leather on cake racks so that it can dry on both sides. Dust the leather lightly with cornstarch or arrowroot powder when all the stickiness has disappeared, then stack the leather in layers with wax paper or aluminum foil between each sheet. Cover the stack with wax paper or aluminum foil and store in a cool, dry place.

Prune Leather

1 gallon pitted prunes
1½ cups water
3 teaspoons almond extract
honey

Place the fruit and water in a large, heavy pot and steam over low heat until the fruit is tender. Drain off the juice and save it for a breakfast drink or punch. (It can be diluted with water half-and-half or more.) Run the pulp through a food mill or sieve. Discard the skins—they are a little too tough to be used in the leather. Add the almond extract and honey to taste. Then spread the pulp on cookie sheets about one-quarter of an inch thick. Dry the pulp as the apricot or peach pulp above until it can be lifted out of the cookie sheet. Then place it on cake racks so that it can dry on both sides. Dust the leather with cornstarch or arrowroot powder when all the stickiness has disappeared, and wrap and store as apricot and peach butter above.

Apple Leather

1 gallon apples
2 to 3 cups apple cider
honey

Peel and core apples, cut them in pieces and run them through a food chopper. Catch the juice which runs from the food chopper and return it to the ground apples. (A blender may also be used. This mashes a limited number of apple pieces at a time, but there is no need to worry about escaping juice.)

Place the ground apples and their juice in a large, heavy pot and add two cups of apple cider. Apples are drier than other fruit and will scorch as they are heated if no liquid is added to the pot. Place the pot over low heat and bring the apples to a boil. Add more cider if needed to prevent the apples from sticking to the bottom of the pot. If the apples are tart, add honey when the mixture looks somewhat clear and is boiling well.

When the mixture reaches the consistency of a very thick apple butter, remove it from the heat and spread the pulp on cookie sheets about one-quarter of an inch thick. Dry this apple pulp just as the other pulps were dried until it can be lifted from the cookie sheets. Then place it on cake racks so that it can dry on both sides. Dust the leather with cornstarch or arrowroot powder when all the stickiness has disappeared, and wrap and store the apple leather as the other leathers above.

HERBS

Harvesting and Drying Herb Leaves

Leaves, seeds, flowers, and even roots of many plants can be collected and dried in the time-honored way, to add distinctive flavor and aroma to many foods all year–round.

Herb leaves are cut when the plant's stock of essential oils is at its highest. In the leafy herbs—basil, savory, chervil, and marjoram—this occurs just before blossoming time. Lemon balm, basil, parsley, rosemary, and sage can be cut as many as four times during the outdoor growing season. Cutting should be done on the morning of a day that promises to be hot and dry. As soon as the dew is off the plants, snip off the top growth—perhaps six inches of stem below the flower buds.

If the leaves are clean, do not wash them; some of the oils will be lost in the rinsing process. If the leaves are dusty or have been thickly mulched, however, wash them briefly under cold water. Shake off any excess water and hang the herbs, tied in small bunches, in the sun, just until the water evaporates from them.

Before the sun starts to broil them, take them in and hang them in a warm, dry place which is well-ventilated and free from strong light. Traditionally, herbs were hung above the mantles of kitchen fireplaces or in attics. Herbs are tied and hung leaves down so that the essential oils in the stems will flow into the leaves. To prevent dust from accumulating on the drying leaves, place a brown paper bag with many holes punched in it for circulation around the leaves. Gather the bag together at the stems and close it with string, a rubber band, or wire tie. A paper bag will also shade the leaves from direct light which would otherwise darken the leaves unnecessarily. Sage, savory, oregano, basil, marjoram, mint, lemon balm, and horehound are best dried in this fashion.

The leaves from thyme, parsley, lemon verbena, rosemary, and chervil may be removed from the stems and spread in a single layer to dry on trays similar to those used for drying fruits and vegetables. Be certain, though, that the screen or netting on the tray is very fine so that no leaves can fall through its openings.

Leaves are most flavorful if they can be dried in three or four days, but conditions are not always right for such quick drying. If the leaves are not entirely dry in two weeks, they should be placed in an oven heated to 100° F. and kept there until they will crumble into dust when rubbed between the palms of the hands. When thoroughly dry, remove leaves from stems.

Traditionally, bunches of fresh herbs were hung by the hearth where the fire's heat would gently dry them. Herbs are hung upside down so that the aromatic oils in the stems will flow into the leaves as they dry.

Harvesting and Drying Herb Seeds

To harvest seeds, gather anise, coriander, cumin, caraway, dill, and fennel plants when the seed pods or heads have changed color, but before they begin to shatter. You may spread the pods one layer thick on a tray just as you do the leaves. When they seem thoroughly dry, rub the pods between the palms of your hands, and the seeds should fall out easily. You can also dry the seeds by hanging the whole plant upside down inside a paper bag to dry. As the seeds dry and fall from the pods, the paper bag will catch them.

Harvesting and Drying Herb Flowers

Flowers to be used in cookery are cut on the first day they are opened. If petals alone are used, they are removed from the calyx and spread on a tray. Rose petals should have their claws—the narrow white portion at the base of each petal—removed. Flower heads used for tea, like camomile, are dried whole.

Harvesting and Drying Roots

The roots of certain plants, particularly angelica, burdock, comfrey, ginseng, ginger, and sassafras, are highly aromatic and can be dried and cut for candy, teas, and cold beverages. These roots are much thicker than the delicate seeds, flowers, and leaves of other plants, and the drying process is a long one.

So as not to injure the plant, roots should be dug out and cut off during the plant's dormant stage, when there is sufficient food stored in the plant cells. This is usually during the fall and winter months. Cut off tender roots— never more than a few from each plant. Scrub them with a vegetable brush to remove all dirt. If the roots are thin, they may be left whole, but for quicker drying, thick roots should be sliced lengthwise. Drying is best done in a low oven. It will require several hours to complete the drying process.

Storing the Herbs

Leaves may be crushed before being stored away, but they retain their oils better if they are kept whole and are crushed right before they are used. Seeds should not be ground ahead of time because they deteriorate quickly when the seed shell is broken. If the seeds are kept whole and stored properly, they will keep several years. Store leaves, seeds, flowers, and roots in tightly sealed jars in a warm place for about a week. At the end of that time, examine the jars. If there is moisture on the inside of the glass or under the lid, remove the contents and spread it out for further drying. Checking the jars is especially important if you are storing dried roots because it is difficult to know when they are completely dry. If further drying is not done, there is a strong chance that mold will develop, and the leaves, seeds, flowers, or roots will not be fit to use. Do not store herbs in cardboard or paper containers, because these materials absorb the oils and leave the dried herbs tasteless. Ideally, herbs should be stored in a cool place, out of strong light, either in dark glass jars, in tins, or behind cabinet doors.

Storing herbs in dried form is so popular because it is a simple means of preservation which produces a product that can be conveniently used in the kitchen. Dried herbs can be taken from their jars just as they are needed, to be mixed with foods while they are cooking or just before they are served. However, the fresh quality of the just-picked herb is lost in the drying process. This is particularly true of chervil, borage, burnet, chives, parsley, and basil. For this reason, some people prefer to preserve some of their herbs by other methods, like freezing and preserving in vinegar. Both of these methods are discussed in other parts of this book (see sections on freezing vegetables and making vinegar), should you be interested in reading about them.

While herbs are usually stored separately, there are advantages to storing some herbs in combination. During storage, the distinct aromas and flavors of the different herbs are given a chance to blend together and form delightful herb mixtures that can be sprinkled on foods or tied in cheesecloth and plunged in soups and stews while they are simmering. Below you'll find some recipes for popular herb combinations. It is important that you do not mix fresh herbs with dried herbs when measuring out for recipes unless you remember that one tablespoon of fresh herbs equals one-half teaspoon of dried herbs or one-quarter teaspoon of dried powdered herbs.

Fines Herbes

Fines herbes is a mixture of finely ground fresh or dried herbs that is sprinkled on food in the last few minutes of cooking or added to the food just before it is placed on the dinner table. Fines herbes can be made from equal parts of any of the following herbs: basil, celery seed, chevil, chives, marjoram, mint, sweet savory, parsley, sage, tarragon, and thyme. Fines herbs are usually served in or on sauces, soups, cheese dishes, and egg entrees.

Bouquets Garnis

Bouquets garnis is a mixture of dried herbs which are tied in a cheesecloth sack or put in a spice ball and plunged into a stew or soup to be left in the food while it is cooking. The sack and the herbs are removed when the stew or soup is taken from the heat. Bouquets garnis can be made from almost any spices, but we recommend the combinations below for starters.

For meat and vegetable soups and stews, mix together:

1 part thyme
2 parts rosemary
1 part sage
2 parts savory
2 parts marjoram

For fish stews and soups mix together:

½ part dill
1 part basil
¼ part oregano
1 part lemon balm
½ part thyme
1 part savory

Herbed butter, which is excellent on roasts, steaks, and fish and good for frying eggs and omelets, can be made by blending one-quarter cup soft unsalted butter with one tablespoon fresh or one-half teaspoon dried herbs. A few drops of lemon juice may be added while blending to bring out the flavor of the mixture.

RECIPES

Dry Roasted Soybeans

1 cup dried soybeans
3 cups water
2 teaspoons instant vegetable salt
1 teaspoon salt

Wash soybeans and remove any foreign particles. Add 3 cups of water to soybeans and soak overnight in refrigerator.

Pour soybeans and liquid used for soaking in a heavy saucepan. Place over medium heat and bring to a boil; lower heat and add instant vegetable salt, cover, and simmer 1 hour. (If desired, ½ teaspoon oil may be added to soybeans to keep mixture from boiling over.)

Remove saucepan from heat and drain soybeans thoroughly; pour into a shallow pan and bake in a preheated 350° F. oven for 45 minutes to 1 hour or until soybeans are brown. Remove from oven and sprinkle with sea salt while warm. Yield: 1½ to 2 cups.

Swedish Soybean Soup

2 cups dried soybeans
water to cover
1 medium smoked ham hock or meaty ham bone
leaves
1 cup chopped onions
3 medium raw turnips, diced
¼ cup chopped parsley

3 quarts cold water
2 teaspoons salt
½ teaspoon paprika
1 cup chopped celery, with
⅛ teaspoon cayenne pepper
1 cup tomato puree or canned tomatoes
chopped parsley for garnish

Wash soybeans and discard beans with imperfections. Cover soybeans with water and place in refrigerator, covered, overnight

The following day, place soaked soybeans in a large, heavy soup kettle. (Be sure to use a large enough pot and leave partially uncovered, so as to avoid soybeans cooking over. This can happen very easily). Add ham hock or meaty ham bone and three quarts of cold water. Place uncovered, over medium heat and bring to a boil, removing any foam from surface as it accumulates. Reduce heat; add 2 teaspoons salt and paprika. Cover partially and allow to simmer for 3 hours, stirring occasionally.

Add the chopped celery, onions, turnips, parsley, cayenne, and tomato puree or canned tomatoes. Cover partially and allow to simmer for another hour or until soybeans are tender. Continue to stir occasionally while cooking. Taste and correct seasoning. Garnish with chopped parsley. Yield: Approximately 3 quarts.

Bean and Corn Soup

1 cup dried white beans
4 cups water or milk
1 teaspoon turmeric
1 onion, chopped
2 cups corn
1 teaspoon cumin
2 cloves garlic

Cook beans with onion, garlic, and spices in simmering liquid. When tender, puree, then add corn and salt to taste. Reheat and serve.

(East) Indian-Style Blackeye Beans

1 cup dried blackeye beans
1 chopped onion
2 tablespoons coconut
¼ teaspoon turmeric
1 teaspoon honey
¼ teaspoon cumin
3 cups water

Add ingredients to boiling water and simmer until liquid is absorbed and beans are tender. Salt to taste.

Black Bean Soup

1 cup dried black beans
4 cups water
3 bay leaves
2 garlic cloves
¼ teaspoon dry mustard
1½ teaspoons chili powder

4 cloves
2 onions, chopped
sea salt to taste

Add ingredients to boiling water and simmer until tender. Puree. Salt to taste and serve.

Wine Red Kidneys

1 cup dried kidney beans
1 onion chopped
½ teaspoon salt
2½ cups water
¾ cup wine

Cook beans and onions in boiling water until tender. Mash half the beans, add salt and wine, then simmer 20 minutes more.

Schnitz and Knepp

This is a specialty of the Pennsylvania Dutch. Schnitz means dried apple slices, and knepp means dumplings.

2 cups dried, sweet apples
1¾ pound hock end of ham
1 egg
2 tablespoons butter
½ cup milk
2 cups flour
3 teaspoons baking powder
½ teaspoon salt

Soak the schnitz in water overnight. Next day, scrub and dry the ham, simmer in water for about 3 hours, then add the soaked apple schnitz and the water in which they have been standing. Boil together for another hour while you make the knepp:

Beat the eggs, add the butter and milk. Sift the dry ingredients and add to the first mixture. There should be just enough milk to make a fairly stiff batter. Drop by spoonfuls into the fast boiling stew, cover the kettle tightly, and steam the knepp for 12 – 15 minutes. Serve on a large platter, placing the meat in the center, the schnitz circling it, and the knepp all around the edge. Yield 6 – 8 servings.

Prune Stuffing

3 cups diced whole wheat bread
½ cup melted butter
1 cup dried prunes
1 cup cold water
1 tablespoon honey
2 tablespoons lemon juice
½ teaspoon salt
½ cup almonds

Mix the bread and the melted butter. Place the prunes in a pan with the water, lemon juice, and honey. Bring to a boil, then simmer for 5 minutes. Save any liquid that may remain and mix with the bread. Pit the prunes, chop, and sprinkle with the salt. Chop the almonds and mix all together. Stuff the bird loosely.

Fruit Mousse

1¼ cups heavy cream
¾ cup cooked dried apricots, peaches, or prunes
1 teaspoon gelatin
4 tablespoons hot fruit juice
½ teaspoon pure vanilla extract
⅛ teaspoon salt

Puree cooked fruit in a blender or through a sieve. In a small bowl, dissolve gelatin in hot fruit juice. Add the gelatin liquid to the pureed fruit. Place the mixture in the refrigerator and chill until it is almost set.

Whip the cream with the vanilla and salt and fold it lightly into the gelatin mixture. Pour into serving dish or individual dessert bowls and chill well before serving. Yield: 4 – 6 servings.

Uncooked Fruit Cake

1 cup ground raisins
1 cup chopped dried figs
1 cup chopped dried dates
1 cup other dried fruits, chopped, (as apricots, pears, prunes, peaches, apples)
1 cup honey
1 cup wholewheat bread crumbs
1 cup concentrated fruit juice or nectar
1 cup sunflower seeds (ground)

Mix in order given. Pack into oiled mold, or bread pan. Ripen, covered, in refrigerator several days. Turn out on platter. Serve with whipped, sweetened cottage cheese or yogurt. Yield: 6 – 8 servings.

Apricot-Nut Bars

⅔ cup dried apricots
½ cup soft butter
¼ cup honey
1⅓ cups whole-wheat pastry flour
½ teaspoon double-acting baking powder
¼ teaspoon salt
½ cup honey
2 eggs, well beaten
½ teaspoon vanilla extract
½ cup chopped walnuts

Rinse apricots; cover with water; boil 10 min. Drain; cool; chop. Start heating oven to 350° F. Oil 8″ x 8″ x 2″ pan.

Mix butter, ¼ cup honey and 1 cup pastry flour. Pack into pan. Bake 25 minutes.

Sift ⅓ cup flour, baking powder, and salt. In large bowl, with mixer at low speed, gradually beat honey into eggs; mix in flour mixture, the vanilla. Stir in walnuts and apricots. Spread over baked layer. Bake 30 minutes or until done; cool in pan. Cut into 32 bars.

Apricot-Cranberry Gelatin

(This salad must be made ahead to set.)

In blender:
1 cup hot water
2 tablespoons gelatin
Blend well, add:
½ cup honey
1 teaspoon pure vanilla extract
¼ teaspoon pure almond extract
½ cup oil
1 cup water
1 cup dried apricots
1½ cups fresh cranberries

Add cranberries and blend only until chopped well. Add any extra you want such as nuts, etc. Pour into cake pan and let set. The chopped fruits will rise to the top and the bottom layer is a creamy-like texture. Yield: 4 servings.

Apricot-Prune Souffle

½ pound dried apricots, cooked
¼ pound dried prunes (pitted and cooked)
5 egg whites
5 tablespoons honey
pinch salt

Press cooked apricots and prunes through a sieve into a large bowl. Add honey and mix thoroughly. Beat egg whites and salt until stiff and fold gently into the fruit. Lightly oil a souffle dish and put the fruit mixture into it. Bake in a slow oven (300° F.) until firm, about 45 minutes. Serve immediately. Yield: 4 servings.

Fruit-Nut Sticks

2 cups nuts
1 cup dates or prunes, pitted, raisins, or dried apricots
2 eggs, beaten
1 cup honey

Grind nuts and fruit. Blend with eggs and honey. Shape into sticks. Place on lightly oiled cookie sheet. Bake at 375° F. for about 10 minutes. Yield: 2 dozen sticks.

Baked Peach Whip

¾ cup cooked dried peaches, sieved or put through blender
4 egg whites—beaten stiff
dash salt
3 tablespoons honey

Fold peach puree into egg whites. Add salt and honey. Mix lightly. Pile lightly into 1 quart casserole, lightly oiled. Bake 30 minutes or until firm. Yield: 6 servings.

UNDERGROUND STORAGE

Underground storage is perhaps the easiest method for storing large amounts of food crops, for once the storage area is constructed, there is very little effort and expense involved in storing a good many fruits and most vegetables. But store only sound fruits and vegetables of good quality this way. Diseased or injured ones may be used early in the fall or preserved in some other way by cutting out bad sections and freezing, canning or drying the rest. Harvesting, in most cases, should be delayed as long as possible without danger of freezing. All vegetables and fruits to be stored must be handled with great care to avoid cuts and bruises. Food that needs to be cleaned, usually just the root crops, should be rubbed lightly with a soft cloth or glove or rinsed under gentle running water. (Let excess water evaporate before storing.)

Root crops keep well when left right in the ground where they grew. Cover the rows with mulch to prevent the ground from freezing, and mark them so that you'll know where to place your shovel when you want to dig up some vegetables.

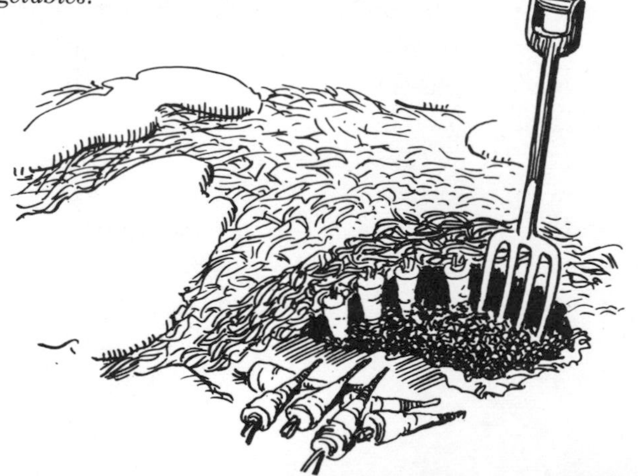

VEGETABLES IN UNDERGROUND STORAGE

Beets, carrots, turnips, and rutabagas–should be harvested in late November, after 30° F. nights. Root crops of this type can be stored by removing the tops–do not wash–and placing them in an area just above freezing, with 95 percent humidity. They can be packed in cans, boxes, or bins, surrounded by straw, or they can be placed in moist sand, or in any outdoor storage pit or root cellar. They may also be left in the garden where they grew. In the middle of the winter you can go out, brush away the snow, dig them up and use them. If you want to make the digging easier, cover the rows with leaves or straw. Make the cover a foot deep, and weight it down with chicken wire or rocks.

Cabbage and chinese cabbage–should be prepared for storage by removing loose outer leaves. If produce is to be wrapped with newspaper, burlap, or some other material, roots and stem should be removed; otherwise leave these in place. Wrapped cabbages should be stored in boxes or bins at a just-above freezing temperature in a cool, damp area. When stems and roots are left on, any of the outdoor storage areas that are made, all or in part, of damp sand or soil is effective. Cabbage emits strong odor during storage that is usually not welcome in the house, so most people prefer to store it in one of the many outdoor storage arrangements.

Celery–is best maintained by pulling the crop. Leave the tops dry, do not wash. The roots should be placed in slightly moist sand or soil, and the plant maintained at 32 to 34° F. To avoid odor contamination, do not store with cabbage or turnips. In areas without very severe winters, celery may be left in the ground, covered with a thick layer of leaves or straw.

Endive or escarole–will keep two or three months in storage if kept at 32 to 34° F. Like celery, endive roots should be kept in slightly moist sand or soil.

Garlic–must be cured before storing. Dry garlic thoroughly, making sure bulbs are not in the direct rays of the sun. For large quantities, garlic may be cured in the garden with its tops covering the bulbs. (Small quantities may be bunched and tied or braided and hung in a well-ventilated cool room to store and dry.) Remove tops and roots with knife or shears leaving one inch of root on the bulb, and store as onions in cool, slightly humid (60 to 75 percent) area.

Kohlrabi—can be stored, after removal of leaves and roots at 32 to 34° F. in an area with about a 95 percent humidity. Root cellars and basement storage rooms are ideal locations.

Onions—must be cured. Leave the vegetable on the ground after pulling for at least two or three days, then place in crates in an open shed for several weeks to complete curing. Remove the tops and store in bins or string bags at temperatures ranging from 33 to 45° F. in an area with about 60 to 75 percent humidity. Attics often prove to be good storage areas.

Parsnips, salsify, and Jerusalem artichokes—can be left in the ground throughout the winter. To make digging easier, cover the rows with about one foot of leaves or straw before the ground has frozen. Jerusalem artichokes are thin skinned and do not keep well once dug up, so dig up no more than a two weeks' supply at a time.

Peas, soybeans and beans—should be shelled and dried. To eliminate fumigation, which is practiced by commercial growers, in order to kill weevils, simply heat the crop in an oven for 30 minutes to an hour at a constant temperature of 135° F. Spread the vegetable in pans for this treatment, and do not let the temperature drop or rise significantly. After drying thoroughly, place in jars or bags for storage. The temperature of the storage area is not important, but it must be dry.

Peppers—if green and mature, they should be picked just before frost. They may be kept for two to three weeks at temperatures between 45 and 50° F. in moderate humidity. Hot varieties of peppers store best if they are dried first and then stored in a cool, dry place. Do not store them in cellars. To dry hot peppers, pull the plants from the ground and hang them up until dried, or harvest the peppers and string them up on a line to dry.

Potatoes—must be stored in the dark. For several months after harvest they can be held in almost any storage location, as this is their normal resting cycle. After this period, temperatures between 34 and 41° F. are necessary to prevent sprouting. The lower temperatures tend to turn starch to sugar and sweeten the vegetable. Only experience with the crop will enable you to determine proper storage in your area. During the storage period, moisture should remain high. Never store with apples.

Pumpkins and winter squash—must also be cured. This helps to toughen their skins so they will keep better. Leave them in the field for two weeks after picking. If weather is near freezing, cure squash in a room with a temperature of about 70° F. for several days. Leave a partial stem on the fruit, and take exceptional care to prevent bruising, storing only the best undamaged produce. After curing, place them gently on shelves, separated from each other, in a 50 to 60° F. dry place. Examine them every few weeks for mold. If you find some, wipe the squash carefully with a cloth made slightly oily with a good vegetable oil. Treated this way, they will keep five to six months.

Sweet potatoes—should be free from injury, and need to be cured before final storage. Lots of air circulation and high temperature over a period of ten days to three weeks is necessary to eliminate excess water, change some starch to sugar, and cause "corking over" of cuts in the skin. After curing, sweet potatoes should be placed in a warm, 50-to-60° F. room which is well ventilated with moderate humidity (up to 75 percent).

Tomatoes—should be harvested just before the first killing frost. Remove tomatoes from plants, wash and allow to dry before storing. Separate green tomatoes from those that show red, and pack green tomatoes no more than two deep in shallow boxes or trays for ripening. Green mature tomatoes will ripen in four to six weeks if held at 55 to 70° F. in moderate humidity.

FRUITS IN UNDERGROUND STORAGE

Many major fruits do not store well for extended periods of time. Of the ones that do—most notably apples and pears—the varieties vary in keeping quality, and it is best to plan to grow good-keeping varieties if you know that you will be storing many of them.

Apples—are of the better-keeping fruit. The Winesap and Yellow Newton are among the best apples, frequently lasting from five to eight months satisfactorily. Next in keeping quality are the Stayman–Winesap, Northern Spy, York Imperial, Arkansas Black Twig, Baldwin, Ben Davis, and Rome Beauty. Normal storage ranges from four to six months with these varieties. Jonathan, McIntosh, Cortland, and Delicious (red or golden) can be kept for shorter periods. There are other factors that influence keeping qualities of different apple varieties such as locality (McIntosh apples grown in New England store better than those grown in the Middle Atlantic States), seasonal conditions, maturity when picked, and length of time between picking and storing.

For long-time storage, apples must be picked at the proper degree of maturity. Apples picked too green are subject to a number of storage disorders, such as scald and bitter pit. If picked beyond maturity, they quickly become overripe in storage.

Good keeping qualities are increased with careful handling to prevent bruising. Storage at between 30 to 32° F. and 85 to 90 percent humidity is preferred for most varieties. Yellow Newton, Rhode Island Greening, and McIntosh are better stored at 36 to 38° F. Wrapping in oiled paper or in shredded paper helps prevent scald, acknowledged to be the most serious disorder, particularly with Cortland and Rhode Island Greening.

Grapes—should be cooled to 50° F. as soon as possible after picking and spread out in single layers. Allow the fruit to remain in this condition until the stems shrivel slightly. Then place the grapes in trays no more than four inches deep in a cellar which is slightly humid and has a temperature of about 40° F. Stored this way, grapes will keep several months.

Oranges—from Florida may be kept eight to ten weeks at 30 to 32° F. with 85 to 90 percent humidity. California oranges can be kept six to eight weeks at 35 to 37° F., but are subject to rind disorders at lower temperatures.

Quinces—will keep two to three months if picked before they are thoroughly ripe and if held in a cool, moist storage area.

Peaches—may be stored several days to two weeks in a cool cellar.

Pears—can be held from eight weeks to several months, depending upon variety. Winter Nelis, Anjou, and Easter Beurre are the most hardy, with Bosc, Kieffer, Bartlett, Comice, and Hardy in the lower range.

Pears should be harvested in a condition that would seem to the amateur to be immature. If allowed to begin to "yellow" on the tree, pears develop hard gritty cells in the flesh. They should be harvested when the dark green of the skin just begins to fade to a yellowish green and the fruit begins to separate more or less readily from the tree.

Pears ordinarily do not ripen satisfactorily at storage temperatures as do apples. For highest eating quality they should be removed from storage while they are still comparatively hard and green, and ripened at room temperature with a high relative humidity. Pears often keep somewhat better in home storage if wrapped in newspaper or other paper. This fruit, like apples, should be stored at a temperature as close to 32° F. as possible and with a high relative humidity, ranging from 85 to 90 percent.

WHERE AND IN WHAT TO STORE

While many of the storage ideas that follow are suitable for fruit storage, fruit should never be stored with potatoes, turnips, or cabbage. The gases released from apples during respiration can cause potatoes to sprout. Cabbage and turnips can transmit their odor to apples and pears. (Wrapping apples and pears in paper or packing them in maple leaves in barrels is recommended if these fruits must be stored with cabbage and turnips, because these materials will prevent absorption of such odors.) Constant air circulation in fruit storage is essential to remove gaseous substances, such as ethylene and volatile esters, which, if they linger, can speed up the ripening process.

Storage Containers

In most storage areas there is a need for small containers. Here are some suggestions:

Wooden boxes, which are orginally designed to store and ship apples and other fruits, make ideal storage units for root cellars or larger storage areas. Interior packing for stuffing between food may be leaves (dry and crisp), hay, straw, string-sphagnum moss, or crumpled burlap.

Helen and Scott Nearing store their Maine garden crops this way with great success. By alternating layers of dried leaves with layers of produce in wooden boxes, they have firm and edible potatoes, apples, rutabagas, carrots, and beets for as long as fifty weeks after they stored them, well into the next growing season.

Pails, baskets, and water-tight barrels are used just as boxes are. Layer packing material and produce alternately, finishing with two inches or more of packing at the top. These containers are used in pit storage areas as well as in larger units.

Metal tins are adaptable for storage, providing you patch-paint raw metal or use galvanized metal to keep rusting at a minimum. Leave them open-topped or cover produce with leaves, sphagnum moss, or straw.

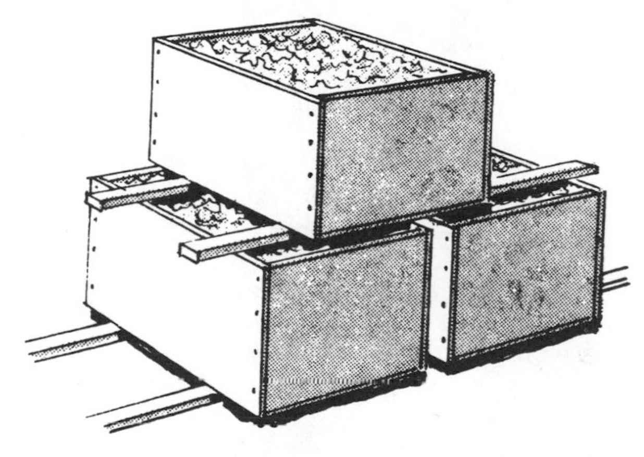

When stacking boxes, place furring strips between them, the floor, and other tiers to permit full air circulation.

Bins are used primarily in larger storage units. They should be constructed for permanent use four inches off the floor. They are good for potato and other root crops.

Orange crates and mesh bags are excellent for onion storage and other foods that need good air circulation.

In the Storage Room

Most vegetables shrivel rapidly unless stored in a moist atmosphere. Shriveling may be prevented: 1. By keeping the air quite moist throughout the storage room. 2. By protecting the vegetables either by wrapping them or putting them in closed containers. 3. By adding moisture directly to the vegetables now and then. If the first method is used, provide a layer of coarse, well washed gravel, three inches thick, on the basement floor. The pebbles should be about equal in size. The shelves and all other equipment are then placed on the gravel. Whenever the cellar air gets too dry (a hygrometer will tell) the gravel is sprinkled with a watering can. Care must be taken, however, that no surplus water is left on the floor.

The wet gravel on the floor will evaporate much more water than on the floor alone, the gravel having a much greater sur-

Watertight barrels and pails are excellent for storing vegetables. Since vegetables need ventilation, cover tops wth a few inches of moist crumpled burlap or string sphagnum moss.

Onions can be safely stored in crates or mesh bags, providing they are placed in a cool, dry place, slightly above the freezing point.

face area. This sprinkling has to be repeated when the relative humidity of the air falls below 80 percent, and for this reason a watering can should always be kept in the cellar. It must be said that one sprinkling usually goes a long way, especially when the cellar is filled to capacity. Garden-fresh produce releases a considerable amount of humidity, any surplus of which can usually be removed with moderate ventilation. However, harsh and prolonged ventilation, open doors, insufficiently insulated walls and doors or a sudden change of outdoor humidity can disturb this balance and make water sprinkling necessary. Only those vegetables which require moist conditions may be stored successfully this way. Vegetables like onions, pumpkins and squash must be stored in some other location where the air is drier. Moist storage rooms are not well suited for canned foods because can lids and metal cans or other metal containers will rust readily.

A dry storage room is more satisfactory for canned foods and also vegetables which prefer a dry place. Root crops, like carrots, which shrivel easily may be stored in a dry room by adding water directly to the vegetables when needed or by placing them in closed containers. Large crocks, metal cans, tight wooden boxes, and barrels are all suitable. Closed containers should be clean, dry and lined with paper before the vegetables are packed; a layer of paper may also be placed between each layer of vegetables. If stored in this way, sand or other materials are not needed to prevent shriveling.

Those who prevent excessive shriveling by adding water directly to the vegetables, commonly store such crops as carrots in crates, boxes, or baskets which are kept covered with burlap or a piece of old rug or carpet. Water is added by sprinkling as needed, and the covering itself is kept moist.

Regardless of the method, stored vegetables should be carefully watched to avoid loss from decay, growth, or excessive shriveling. Decaying vegetables should be taken out as soon as noticed. If vegetables start to grow, the temperature is too high. Vegetables which begin to shrivel a great deal should be wrapped, placed in closed containers, or sprinkled with water. In a moist storage room extra moisture may be provided by sprinkling the floor.

Cleanliness is one of the first things to be observed in storage cellars. Walls and ceiling should be whitewashed and the floor must be kept clean at all times. Dead leaves, stalks, and the like must be removed from under shelves and planking. Spring cleaning is obligatory and it is best done during dry summer days. Shelves, crates, baskets, and other containers must be brushed in the open air or with all windows open.

Basement Storage Ideas

The cellars in old houses didn't have dirt floors just because they were cheap and easy to construct. These dirt floors made cellars excellent food storage areas; they helped to keep food cool and moist. These old root cellars have long been used for storage in the colder parts of our country, and some houses without central heating on farms and homesteads are now being built with dirt floor cellars for just this reason. These cellars usually have an outside entrance which can be opened to ventilate the cellars and regulate the temperature inside. Many have insulating material on the ceiling to prevent cold air in the cellar from chilling the whole house.

Centrally heated homes with concrete floor basements are generally too warm to be used just as they are for food storage. But, with a little ingenuity and a minimal expense, part of just about any basement can be converted into a storage room that will pay for itself in one year in pared-down food bills if you apply basic storage principles, discussed below.

Essentially the storage room is a place where temperature and humidity are held to the proper level for keeping produce. This means lower-than-usual household or basement temperatures, ranging from 30 to 40° F.

With few exceptions, the most desirable temperature is at or very near 32° F.—the freezing point of water. Except for potatoes, vegetables are not injured at this temperature. It is difficult, however, to keep the temperature as low as 32° F. without danger of it going low enough to cause actual freezing during exceedingly cold weather. It is suggested, therefore, that the storage room temperature be kept between 35 and 40° F. Such temperatures cannot be reached and kept except in a room separated from the rest of the basement, reasonably well-insulated, and having adequate ventilation.

The size of the basement storage room will vary with the space available and the family needs; eight by ten feet is suggested for most families who plan to store both vegetables and other foods in the same room, a room this size will hold sixty bushels of produce. A storage room, if properly constructed and managed, will be suitable for nearly all foods commonly preserved. Where practical, the storage room should be located either in the northeast or northwest corner of the basement and away from the chimney and heating pipes.

At least one wall having outside exposure should be used, preferably the one with the least sunlight, on the north side and with a window that is easily reached. The other walls can be made of wood which will do a good job of keeping the storage area cool—providing their construction is tight.

Two types of insulating material can be used: board or loose fill. It is important to keep them dry or their insulating properties will be reduced. So moisture-vapor barriers are used, inside and out, such as damp-proof paper, tar, asphalt.

Board insulation can be nailed to the walls and ceiling. Two thicknesses should be used to prevent leakage through joints.

Loose-fill insulation includes planer shavings, cork dust and minerals. Stagger the studs so they are exposed on one surface only. Sheathing and damp-proofing should be done when the wall space is being filled.

Planer shavings are very satisfactory insulation for side walls and ceiling; they are dry and do not tend to settle. Add hydrated lime to the shavings, from twenty to forty pounds per cubic yard, to help keep the shavings dry and as a repellent to vermin and rodents. Tamp the shavings as they are filled between studs until a density of seven to nine pounds per cubic foot is obtained.

Loose-fill insulation is not used where floor insulating is needed. Use board-type insulation—wood-fiber insulating wall-board to floor insulation. (The floor to side wall jointing is the same as for wall corners.) The insulating board is then mopped with hot tar, and cement or other flooring is laid on top. Where chinks occur at rough wall or floor surfaces, caulking compound, which fills the space and protects the exposed insulating material from dampness, is used.

Temperature of the surrounding basement will determine thickness of the walls. A prevailing temperature of 60° F. outside the storage room calls for the following thickness of insulating materials:

3 inches: wood-fiber insulating board, cork board, granular cork, fibrous rock, rock wool.

4 to 5 inches: planer shavings.

6 inches: compressed peat.

8 to 9 inches: white pine and soft woods.

Lower basement temperatures will require less insulation.

The walls to close off a corner of the basement are easily constructed with a two-by-four-inch framework, sheathing on both sides and three-inch insulation batts between the studding. Leave an opening in one wall for a door. This may be framed with two-by-two-inch studs, faced on each side with quarter-inch plywood, and the center filled with insulation. Fit the door tightly and secure it with a type of latch that holds it firmly closed.

To insulate the ceiling of the storage space, sheathe underneath the ceiling joists and apply four inches or more of insulation between the joists, extending the insulation out over the walls of the storage space.

Ability to maintain a desirable temperature range of 35 to 40° F. in the basement storage room depends largely on outside weather conditions. In both early fall and late spring, day temperatures are likely to be higher than those desired in the storage area. Therefore, it is important during these times to keep windows closed. As a general rule, windows should be opened whenever the outside temperature is lower than that in the storage room and the inside temperature is above 40° F. When the temperature in the storage room drops to 35° F., windows should be closed. Place one reliable thermometer inside and another outside a window for guidance.

Light must be excluded from stored vegetables and fruits, so cover the windows with opaque material. Wide, wooden louvres fitted to the outside of the window frame aid in excluding light if the window is opened for ventilation in the daytime. Cover louvres, or open windows if louvres are not used, with copper screening to keep out insects and animals.

A small, but simple and inexpensive storage area can be made by utilizing the cellar steps of an outside cellar entrance. Install an inside door to keep out basement heat at the bottom of the steps. If you want to create an even large storage area, build inward into the basement, but take care to insulate extra wall space. Temperatures in the stairwell will go down as you go up the steps, and a little experimenting will help you determine the best levels for the different crops you are storing. If the air is too dry, set pans of water at the warmest level for extra humidity.

In a pinch, window area wells can hold bushels of food over part or all of the winter. Because they are adjacent to the house and below ground level, the temperatures inside them should remain fairly constant throughout the winter. Cover the wells with screening and wood. To raise the temperature in very cold weather, open basement window and allow some house heat to enter. When the temperature in the window well gets too high, remove wood from the top of the well to permit the cold, outside air to cool the area. If basement windows open inward or are the sliding type, access can be convenient and simple during the cold winter months.

Down in Myerstown, Pennsylvania, Enos Hess, with the help of his county farm agent, has constructed an extensive and unique storage facility in the form of a modern "old-fashioned root cellar." He has constructed a large garage-type building attached to the house in which he stores packaged foods. At the outside wall of the structure, and near the garage door, a door opens onto a stairway and conveyor belt which leads downward for about twelve feet.

Behind the door at the base of the steps is a 12-by-18-foot room with 12-inch-thick concrete block walls and a 10-inch thick reinforced concrete ceiling, covered by four solid feet of earth. The cellar is built over a natural spring which is covered by a four-inch-thick concrete slab. Hydrostatic pressure forces the cool water up though the slab, wetting the surface of the concrete floor. Wood slats lie over the concrete, and any excess water is pumped out of the cellar automatically.

Air circulation is most effective. Intake air entering through a 12-inch-round concrete overhead flue is directed out at the floor through an adjustable grill. The air is cooled as it passes over the water and flows through the room.

Asked why he installed a concrete floor when the water condition was as indicated, Enos explained there could be a muddy condition if he hadn't used concrete. He is so well-satisfied that when he builds the adjacent 12-by-30-foot room this summer, he will not change this proven method one bit.

The flat ceiling created one problem he had not anticipated which will be solved in his new room. Condensation on the ceiling causes dripping all over the room. A tent-like installation of plastic sheeting sloping to either side of the room catches the water and directs it to the walls, and thence to the floor.

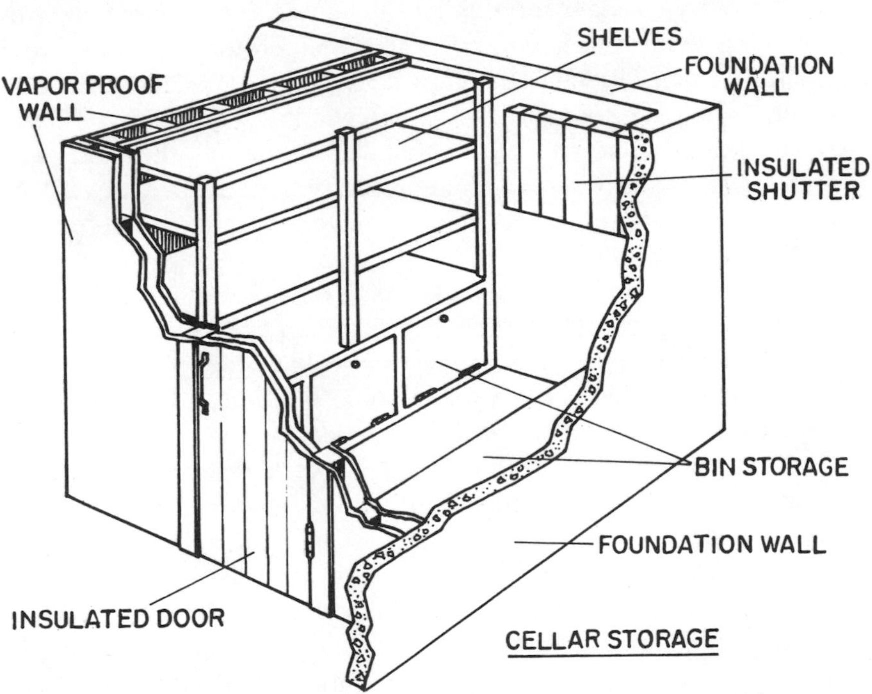

The basement often provides needed storage space for fruit and vegetable storage, but too much heat and too little humidity frequently cancel out its usefulness. A cellar storage room can be constructed rather easily by the home handyman. See the following page for plans.

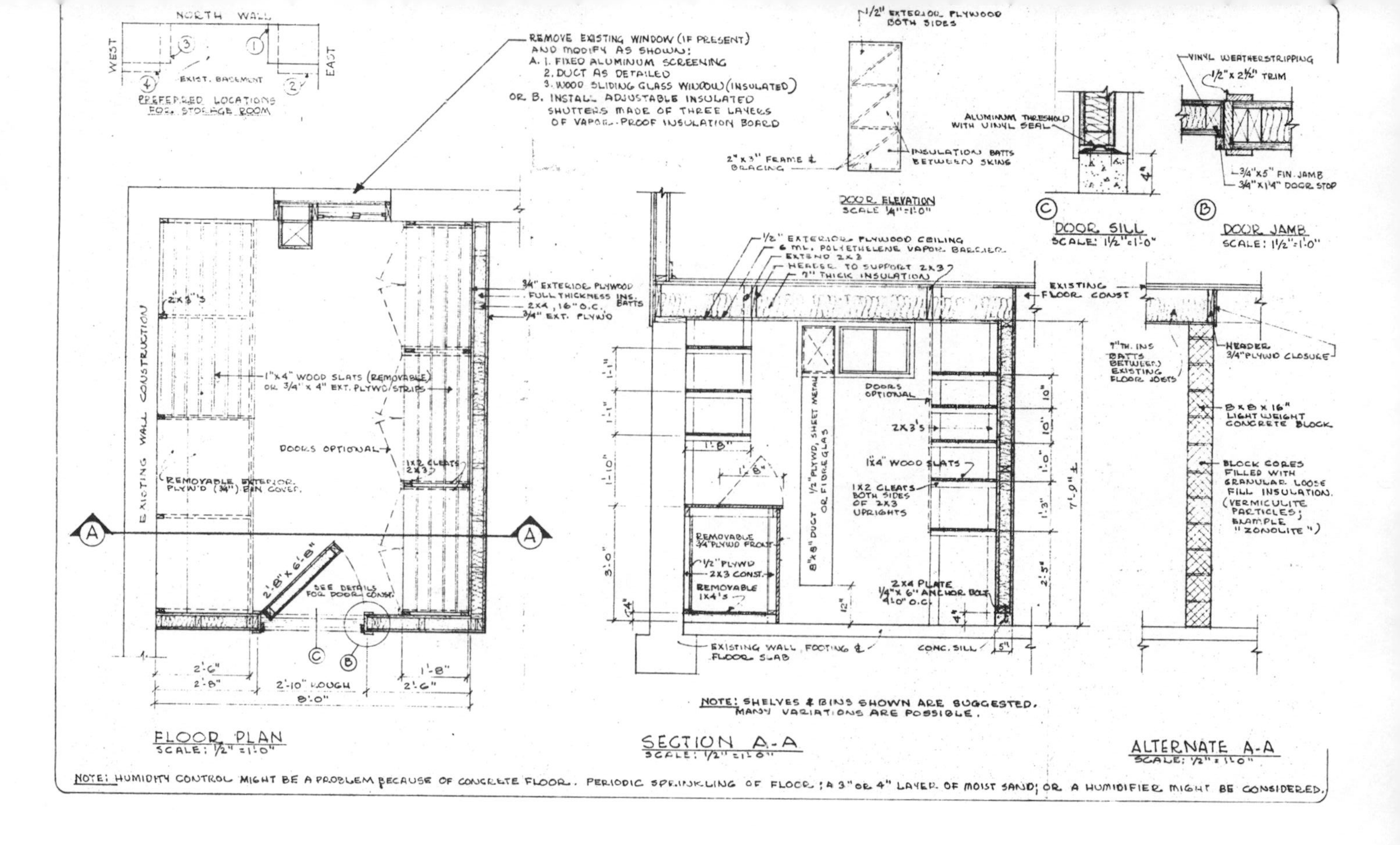
NORTH WALL
WEST
EAST
EXIST. BASEMENT
PREFERRED LOCATIONS FOR STORAGE ROOM
REMOVE EXISTING WINDOW (IF PRESENT) AND MODIFY AS SHOWN:
A. 1. FIXED ALUMINUM SCREENING
2. DUCT AS DETAILED
3. WOOD SLIDING GLASS WINDOW (INSULATED)
OR B. INSTALL ADJUSTABLE INSULATED SHUTTERS MADE OF THREE LAYERS OF VAPOR-PROOF INSULATION BOARD
1/2" EXTERIOR PLYWOOD BOTH SIDES
2" x 3" FRAME & BRACING
INSULATION BATTS BETWEEN SKINS
DOOR ELEVATION
SCALE 1/4" = 1'-0"
ALUMINUM THRESHOLD WITH VINYL SEAL
DOOR SILL
SCALE: 1 1/2" = 1'-0"
VINYL WEATHERSTRIPPING
1/2" x 2 1/2" TRIM
3/4" x 5" FIN. JAMB
3/4" x 1 1/4" DOOR STOP
DOOR JAMB
SCALE: 1 1/2" = 1'-0"
3/4" EXTERIOR PLYWOOD
FULL THICKNESS INS. BATTS
2X4, 16" O.C.
3/4" EXT. PLYWD
2X3"'S
1"x4" WOOD SLATS (REMOVABLE) OR 3/4" x 4" EXT. PLYWD STRIPS
DOORS OPTIONAL
1X2 CLEATS 2X3
(REMOVABLE EXTERIOR PLYW'D (3/4") BIN COVER
EXISTING WALL CONSTRUCTION
2'-8" x 6'-8"
SEE DETAILS FOR DOOR CONST.
2'-6"
2'-8"
2'-10" ROUGH
1'-8"
2'-6"
8'-0"
FLOOR PLAN
SCALE: 1/2" = 1'-0"
1/2" EXTERIOR PLYWOOD CEILING
6 ML. POLYETHELENE VAPOR BARRIER
EXTEND 2X3
HEADER TO SUPPORT 2X3
7" THICK INSULATION
EXISTING FLOOR CONST
1'-1"
1'-1"
1'-10"
3'-0"
4"
1'-8"
1'-8"
REMOVABLE 3/4" PLYWD FRONT
1/2" PLYWD
2X3 CONST.
REMOVABLE 1X4'S
8"x8" DUCT 1/2" PLYWD, SHEET METAL OR FIBREGLAS
12"
DOORS OPTIONAL
2X3'S
1X4" WOOD SLATS
1X2 CLEATS BOTH SIDES OF 2X3 UPRIGHTS
2X4 PLATE
1/4" x 6" ANCHOR BOLT 4'-0" O.C.
10"
10"
1'-0"
1'-3"
2'-5"
7'-9" ±
EXISTING WALL FOOTING & FLOOR SLAB
CONC. SILL
5"
NOTE: SHELVES & BINS SHOWN ARE SUGGESTED. MANY VARIATIONS ARE POSSIBLE.
SECTION A-A
SCALE: 1/2" = 1'-0"
7" TH. INS BATTS BETWEEN EXISTING FLOOR JOISTS
HEADER
3/4" PLYWD CLOSURE
8 x 8 x 16" LIGHT WEIGHT CONCRETE BLOCK
BLOCK CORES FILLED WITH GRANULAR LOOSE FILL INSULATION. (VERMICULITE PARTICLES; EXAMPLE "ZONOLITE")
ALTERNATE A-A
SCALE: 1/2" = 1'-0"
NOTE: HUMIDITY CONTROL MIGHT BE A PROBLEM BECAUSE OF CONCRETE FLOOR. PERIODIC SPRINKLING OF FLOOR; A 3" OR 4" LAYER OF MOIST SAND; OR A HUMIDIFIER MIGHT BE CONSIDERED.

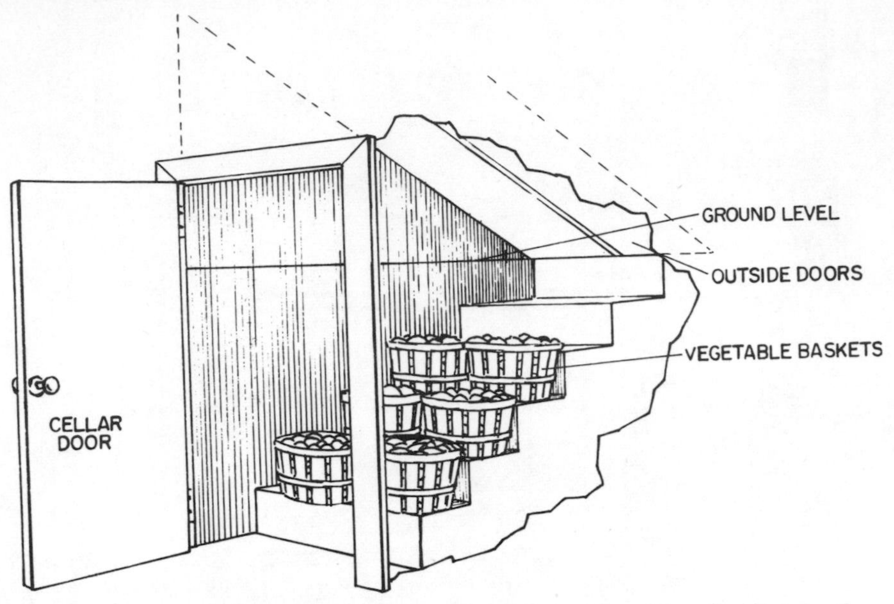

If your house has a sloping cellar door with an outside stairwell into the basement, you can adapt this area into storage with relative ease and little expense.

Enos Hess has very successfully stored potatoes, beets and carrots in open boxes and celery and cabbage in plastic bags well into spring of the following year.

A Homemade Storage Unit

Do you know of a supermarket going out of business or remodeling? With a little luck, some effort, time, and determination, you may be able to have a six-cubic-foot cold storage unit sitting in your house for the meager price of twenty-five dollars. Grace and Tim Lefever of Sonnewald Farm near Spring Grove, Pennsylvania do. Although they had to chop up and cart away a six-inch concrete slab in the process, the Lefevers got an insulated ice-maker box from a supermarket for free. They put it in an unheated area in their house and built a new floor for the box by placing five-inch-deep styrofoam between wooden supports that were cut to a uniform slope to the drain in the center. This understructure was covered with plywood and finally faced off with a sheet metal floor. Although apples, beets, carrots, green beans, onions, potatoes, and turnips have been successfully stored

Enos Hess has successfully stored a variety of crops in open boxes in his storage room, which is something of a modern "old-fashioned root cellar." The entire chamber is built of wooden siding over 12-inch-thick concrete block walls and a 10-inch-thick reinforced concrete ceiling—all under a solid 4 feet of earth.

over the winter without the box being closed, a door will eventually be added.

Storage in an Old Truck Body

John Keck, of the Rodale Organic Experimental Farm, didn't have a cellar to work with, so be constructed an outdoor, underground storage area from an old steel "REA" Express truck body. He supported the roof of the truck body with pipe columns and light beams, imbedded it into a bank and sodded it down with earth. (He found that the steeper the bank and the more earth coverage, the better.) A small stairway down to the truck acts as a door entry. Ventilation is provided through the roof in two places.

Cost of the materials, including the truck body, was less than $150. Complete details follow on the next page.

A Concrete Storage Room

Root cellars, like the one below, can be made from scratch by making a concrete structure and then burying it below

Organic farm manager John Keck tests the hand-fitted entrance he made for the earth-covered storage shed which was originally an auto truck delivery body.

ground level. For details, see the plans and material specifications that follow.

But storage areas need not be so intricate. Many gardeners and small-scale farmers have found that the ground itself can protect vegetables and fruits from winter's freezing temperatures. They have taken advantage of the ground's insulating qualities to make small, efficient storage space for their garden products.

Soil-Pit Storage

The glacial plain of northwestern Pennsylvania provides excellent conditions—moist soil with a winter temperature around 52° F.—for soil-pit storage like the kind Jane Preston devised:

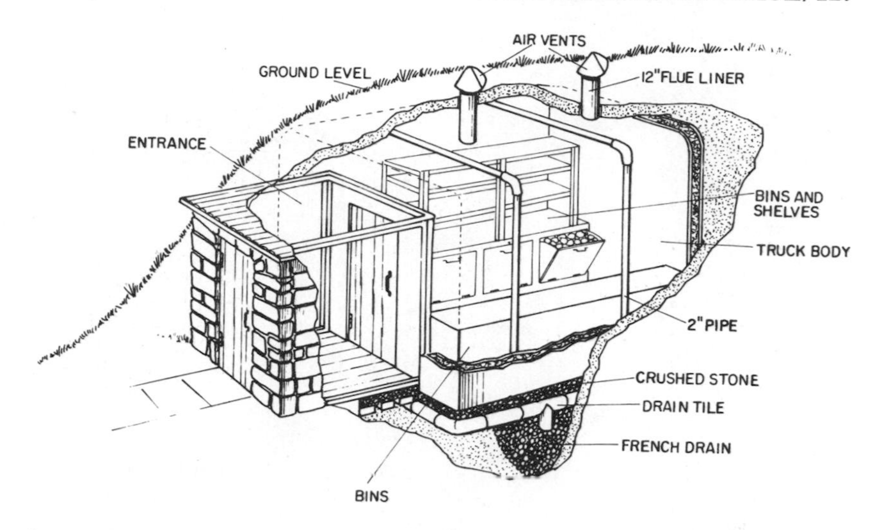

The truck body rests on a bed of crushed stone and is covered with earth for insulation. An entrance made from concrete blocks makes the storage area easily accessible.

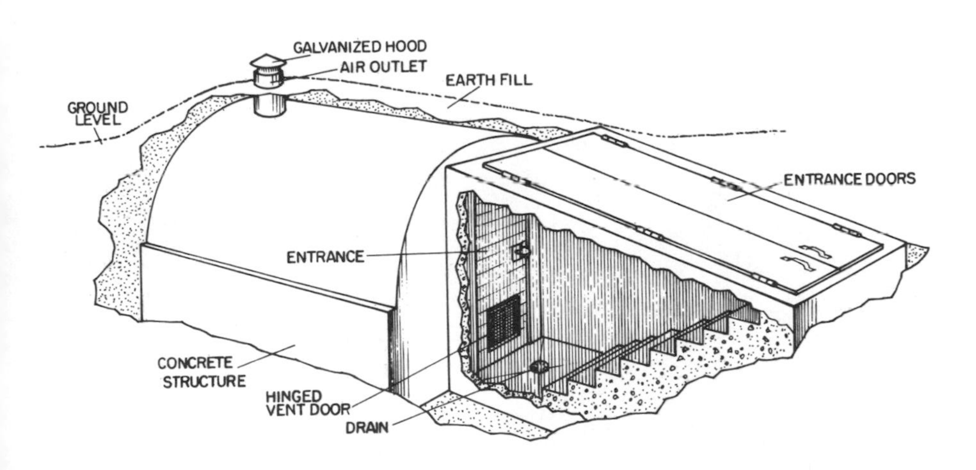

For proper insulation, make the depth of the earth covering this storage structure 12 inches deeper than the local frost line depth. This means that if the frost line is 30 inches deep, the earth cover should be 42 inches in depth.

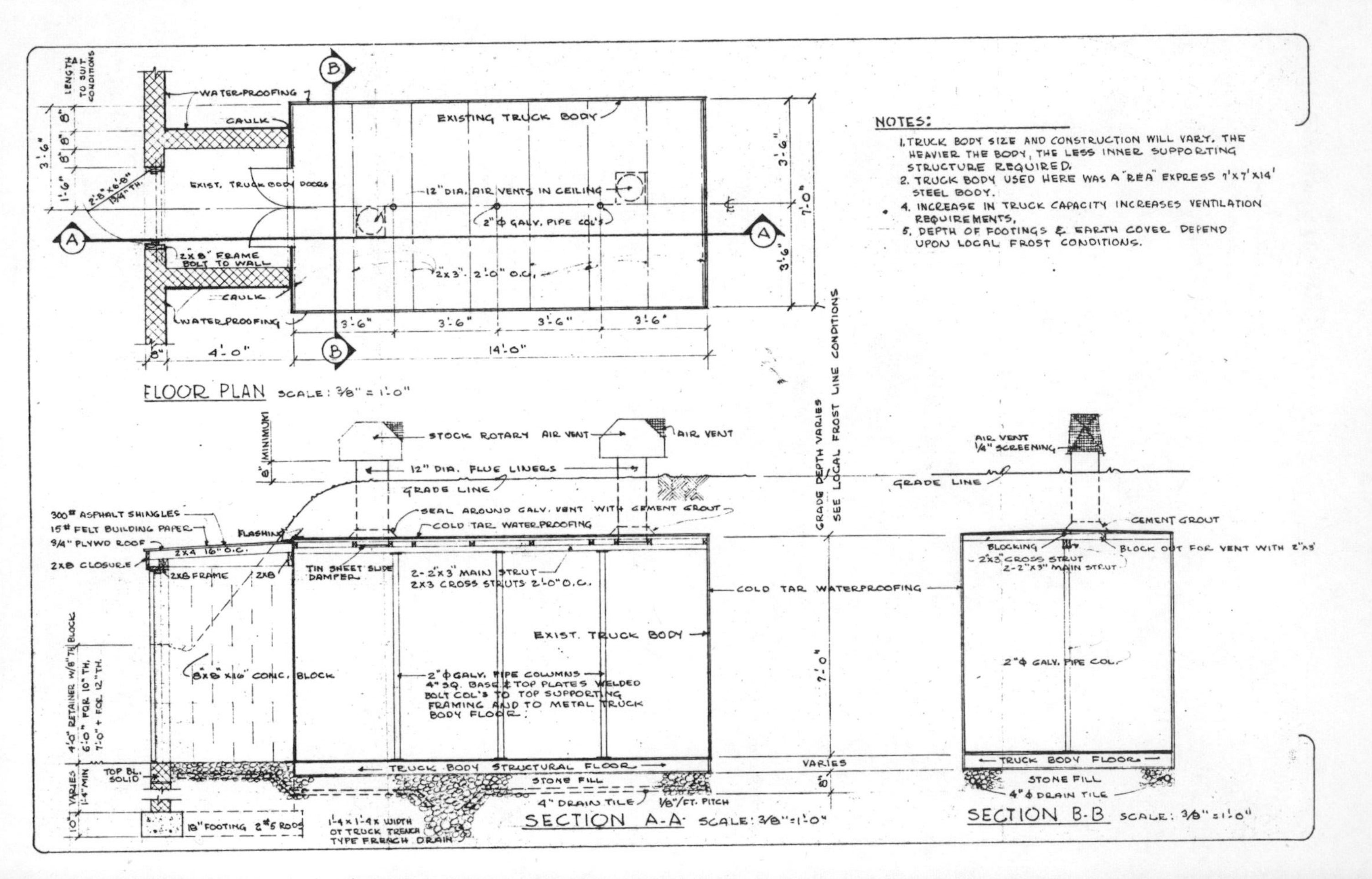

NOTES:
1. TRUCK BODY SIZE AND CONSTRUCTION WILL VARY. THE HEAVIER THE BODY, THE LESS INNER SUPPORTING STRUCTURE REQUIRED.
2. TRUCK BODY USED HERE WAS A "REA" EXPRESS 7'x7'x14' STEEL BODY.
4. INCREASE IN TRUCK CAPACITY INCREASES VENTILATION REQUIREMENTS,
5. DEPTH OF FOOTINGS & EARTH COVER DEPEND UPON LOCAL FROST CONDITIONS.
FLOOR PLAN SCALE: 3/8" = 1'-0"
LENGTH TO SUIT CONDITIONS
WATER-PROOFING
CAULK
EXISTING TRUCK BODY
EXIST. TRUCK BODY DOORS
12" DIA. AIR VENTS IN CEILING
2" ϕ GALV. PIPE COL'S
2'-8"x6'-8" 1 3/4" TH.
2x8" FRAME BOLT TO WALL
2x3". 2'-0" O.C.
CAULK
WATER-PROOFING
3'-6"
3'-6"
3'-6"
3'-6"
3'-6"
3'-6"
7'-0"
1'-6"
8"
4'-0"
14'-0"
SECTION A-A SCALE: 3/8" = 1'-0"
STOCK ROTARY AIR VENT
AIR VENT
12" DIA. FLUE LINERS
GRADE LINE
8" MINIMUM
300# ASPHALT SHINGLES
15# FELT BUILDING PAPER
3/4" PLYWD ROOF
2x8 CLOSURE
2x4 16" O.C.
FLASHING
2x8 FRAME
2x8
SEAL AROUND GALV. VENT WITH CEMENT GROUT
COLD TAR WATERPROOFING
TIN SHEET SLIDE DAMPER
2-2"x3" MAIN STRUT
2x3 CROSS STRUTS 2'-0" O.C.
EXIST. TRUCK BODY
8"x8"x16" CONC. BLOCK
2" ϕ GALV. PIPE COLUMNS
4" SQ. BASE & TOP PLATES WELDED
BOLT COL'S TO TOP SUPPORTING FRAMING AND TO METAL TRUCK BODY FLOOR.
4'-0" RETAINER W/8" TH. BLOCK
6'-0" FOR 10" TH.
7'-0" + FOR 12" TH.
TOP BL. SOLID
VARIES
1'-4" MIN
10"
10" FOOTING 2 #5 RODS
1'-4 x 1'-4 x WIDTH OF TRUCK TRENCH TYPE FRENCH DRAIN
TRUCK BODY STRUCTURAL FLOOR
STONE FILL
4" DRAIN TILE
1/8"/FT. PITCH
GRADE DEPTH VARIES
SEE LOCAL FROST LINE CONDITIONS
COLD TAR WATERPROOFING
7'-0"
VARIES
8"
SECTION B-B SCALE: 3/8" = 1'-0"
AIR VENT
1/4" SCREENING
GRADE LINE
CEMENT GROUT
BLOCKING
2x3" CROSS STRUT
2-2"x3" MAIN STRUT
BLOCK OUT FOR VENT WITH 2"x3"
2" ϕ GALV. PIPE COL.
TRUCK BODY FLOOR
STONE FILL
4" ϕ DRAIN TILE

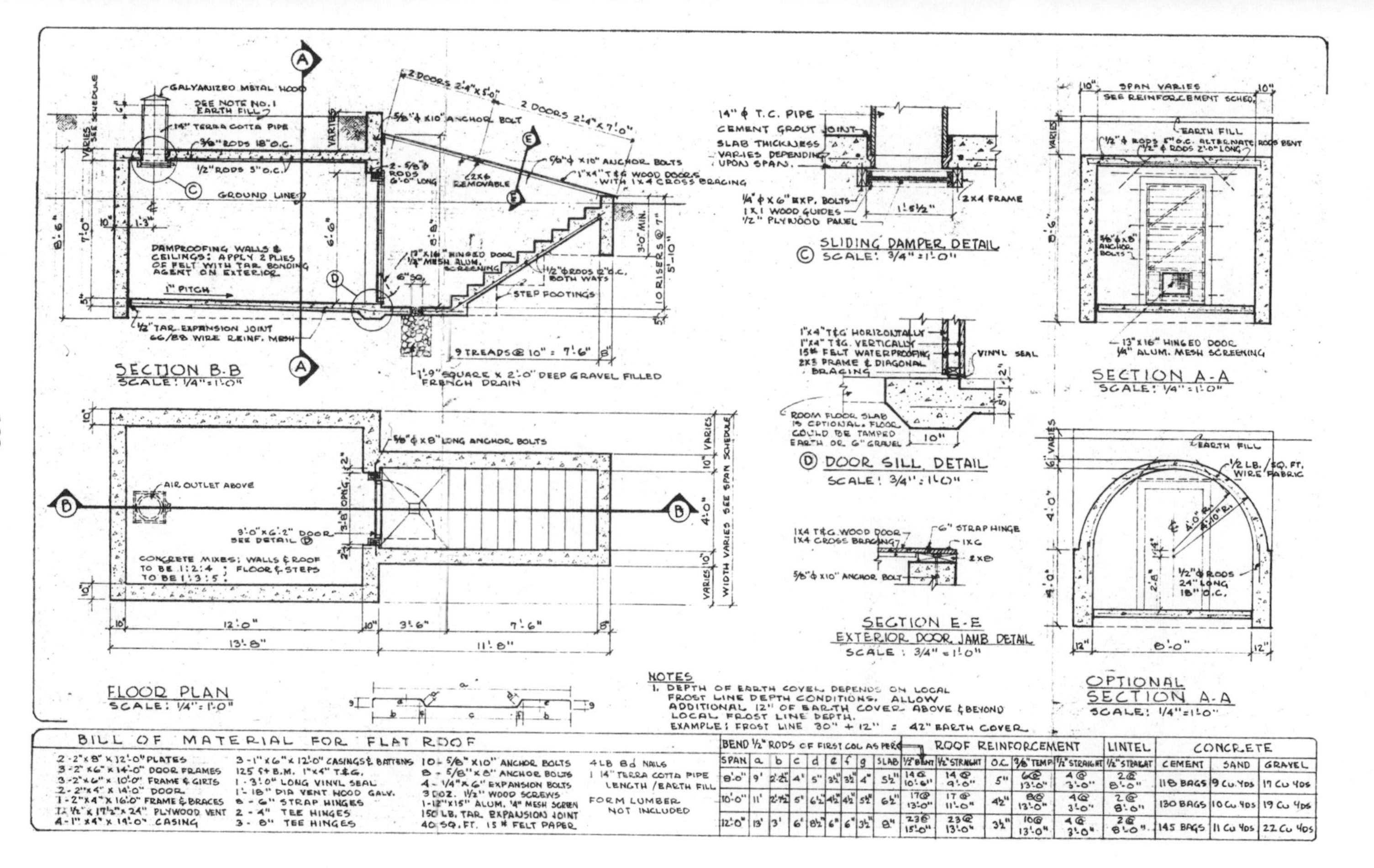

NOTES

1. DEPTH OF EARTH COVER DEPENDS ON LOCAL FROST LINE DEPTH CONDITIONS. ALLOW ADDITIONAL 12" OF EARTH COVER ABOVE & BEYOND LOCAL FROST LINE DEPTH.
EXAMPLE: FROST LINE 30" + 12" = 42" EARTH COVER

BILL OF MATERIAL FOR FLAT ROOF

2-2"x8"x12'-0" PLATES
3-2"x6"x14'-0" DOOR FRAMES
3-2"x6"x10'-0" FRAME & GIRTS
2-2"x4"x14'-0" DOOR
1-2"x4"x16'-0" FRAME & BRACES
1-1/2"x17 1/2"x24" PLYWOOD VENT
4-1"x4"x14'-0" CASING
3-1"x6"x12'-0" CASINGS & BATTENS
125 ft B.M. 1"x4" T&G.
1-3'-0" LONG VINYL SEAL
1-18" DIA VENT HOOD GALV.
8-6" STRAP HINGES
2-4" TEE HINGES
3-8" TEE HINGES
10-5/8"x10" ANCHOR BOLTS
8-5/8"x8" ANCHOR BOLTS
4-1/4"x6" EXPANSION BOLTS
3 DOZ. 1 1/2" WOOD SCREWS
1-12"x15" ALUM. 1/4" MESH SCREEN
150 LB. TAR EXPANSION JOINT
40 SQ. FT. 15 # FELT PAPER
4 LB 8d NAILS
1 14" TERRA COTTA PIPE LENGTH / EARTH FILL
FORM LUMBER NOT INCLUDED

BEND 1/2" RODS OF FIRST COL AS PER									ROOF REINFORCEMENT					LINTEL	CONCRETE		
SPAN	a	b	c	d	e	f	g	SLAB	1/2" BENT	1/2" STRAIGHT	O.C.	3/8" TEMP	1/2" STRAIGHT	1/2" STRAIGHT	CEMENT	SAND	GRAVEL
8'-0"	9'	2'-2 1/2"	4'	5"	3 1/2"	3 1/2"	4"	5 1/2"	14 @ 10'-6"	14 @ 9'-0"	5"	6 @ 13'-0"	4 @ 3'-0"	2 @ 8'-0"	118 BAGS	9 CU. YDS	17 CU YDS
10'-0"	11'	2'-7 1/2"	5'	6 1/2"	4 1/2"	4 1/2"	5 1/2"	6 1/2"	17 @ 13'-0"	17 @ 11'-0"	4 1/2"	8 @ 13'-0"	4 @ 3'-0"	2 @ 8'-0"	130 BAGS	10 CU YDS	19 CU YDS
12'-0"	13'	3'	6'	8 1/2"	6"	6"	3 1/2"	8"	23 @ 15'-0"	23 @ 13'-0"	3 1/2"	10 @ 13'-0"	4 @ 3'-0"	2 @ 8'-0"	145 BAGS	11 CU YDS	22 CU YDS

Our pit is neatly cut out, three feet wide, six feet long, and two feet deep. It is on sloping ground, so that excess water promptly drains away. We lined the inside entirely with ¼-inch hardware cloth, carefully kept tight—as it *must* be—to keep out rodents. The top is finished with a two-by-four-inch frame and a neat wooden lid. (If mice do get in, they are in "clover" and you are in deep trouble!)

At harvest time—late autumn—select carrots, beets, parsnips, potatoes, turnips, and the like. Wash them free of soil, but be careful not to bruise the skin. Put a layer of clean, sharp builder's sand (washed) on the pit bottom. Then place a neat layer of root vegetables. Do not "dump" vegetables into the pit. After the first layer of root crops is in place, cover with a layer of sand, then continue as you fill the pit. A "map" is mighty helpful in keeping account of where different roots are to be found. Finish the pit with a final layer of sand.

Four bales of straw can be laid on the cover, and this then is covered with a plastic sheet to keep off snow. The insulation provided by the straw bales not only helps keep things "warm" in winter (avoiding freezing), but also cool in the warm days of early spring. Good-quality vegetables have been taken from the pit as late as May—even early June. Then it is time to clean everything out and let in sunshine and ventilation over the summer.

Our pit has been in steady use for over ten years. We have had to replace the wooden parts or mend wire on occasion, but that's about all. And we've been able to supply a home-grown table with good varieties of all root vegetables, plus cabbage and celery as well. The secret, if there is one, resides in 100 percent humidity at a very uniform temperature all winter.

Hay-Bale Storage

Out in Ohio, gardener John Krill stores his apples and root crops in on-top-of-the-ground storage areas. Building these areas is quite simple, and the materials used may be turned into mulch

Separate layers of root vegetables by layers of sand in this pit-type storage area to retain moisture and keep the temperature constant.

or compost after the storage season. Here John describes how he built them:

Bales of spoiled hay are commonly available nearly everywhere at low cost. Bales average in size about 42 inches long by 18 inches high and 14 inches wide. I lay two bales end-to-end. About 14 inches across from them, I place two more bales end-to-end paralled to the first two. This is done on a well-drained spot, of course. The ends are closed off by placing a bale across each end. A thin layer of hay is then placed in the bottom of the resulting "box."

Apples, pears, potatoes or other root crops are then placed inside of this hay bale enclosure. Do not dump the produce in. Careless handling will cause bruising which will quickly

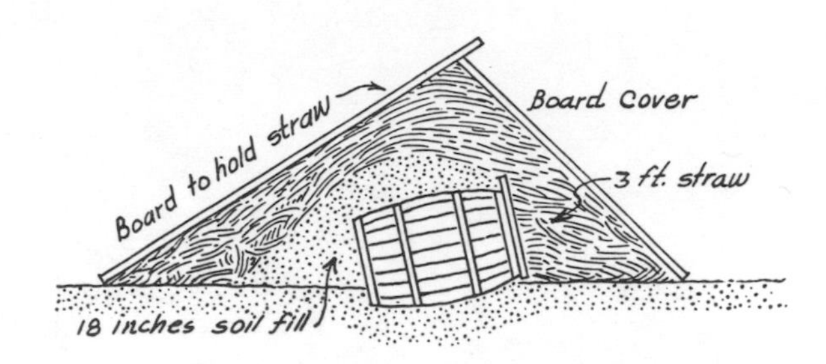

A variation of the Prestons' soil pit storage is the barrel pit, which is made by partially burying a barrel with earth and covering it with straw and used lumber.

cause rot. Place carefully by handfuls until within several inches of the top. Cover top with hay. Place hay bales across opening. Place a stone under each bale on top to keep it from completely sealing the opening.

Wait until freezing weather sets in, then remove the stones so that the top bales may completely seal the opening. This allows for natural ventilation of the stored crop. When unseasonably warm weather comes, lift one bale and place a stone under it to permit ventilation until severe cold again sets in. Then remove stone to again seal the top.

The great thickness of the hay bales make excellent insulation against cold. Enough air seeps through to create ventilation without permitting freezing cold to enter. Dig a shallow trench alongside of the bottom bales to carry away rain.

These hay-bale enclosures are so quickly and simply made that a number of them may be constructed. No soil is heaped up around them as in most other methods of storage. Access to the stored produce is quickly and easily gained. A single top bale is removed to expose the interior. It is easily replaced after removal of a quantity of the contents.

After the contents have been used up, the hay bales are converted either into mulch or added to the compost pile. Thus the bales play two important parts in the scheme of

Top-of-the-ground storage is suitable for most root crops. Here, a rectangle is made from bales of hay. A lid made from additional hay bales covers the food in the center. A stone can be placed under the top bales for ventilation during mild weather and can be removed when freezing weather prevails.

the organic gardener. But the best appreciated part is the lack of hard physical labor and trouble in constructing the hay-bale storage enclosures.

I have kept tomatoes ripening on their vines long after frosts invaded our area. The method is quite simple and highly effective. I stake my tomatoes and the stake plays an important part in the method. All small tomatoes that have no chance of ripening are removed. Only those showing signs of even the faintest blush color are kept on the vine. The vine is then tied as compactly as possible to the stake.

Next, old hay is pushed up around and over each tomato plant like the skirt around the waist and legs of a hula dancer. Keep this hay wrap loose and three or four inches thick, and secure by wrapping with twine around the tomato plant and its stake. The stake makes a support for the entire arrangement. To pick tomatoes, carefully part the hay without pulling it free. Remove the ripened fruit and replace the hay.

Tomatoes kept in this fashion are far more trouble-free than by other methods commonly used. Best of all, the tomato tastes like a garden-ripe fruit. They will slowly ripen in this fashion until freezing weather arrives to stay.

Storing in Garbage Pails

Gardener Elmer L. Onstott, of St. Louis, Missouri, had a problem keeping his late-summer vegetables garden-fresh over the winter because he had no room for basement storage. A good part of his storage problem was solved when he discovered that ten-gallon garbage pails can be converted into storage bins by burying them in sixteen-inch holes, leaving their rims above ground. In addition to being inexpensive and readily accessible, the garbage-pail storage bins are water-and rodent-proof and store easily over the summer as a compact stack of pails.

Here is how Mr. Onstott got his project of year-round fresh vegetables started:

> Late in November or in early December, I wait for several 30° F. nights and then start harvesting my carrots, beets, and turnips. All I have to do is pull them, shake off the loose dirt, cut the tops off, and place them in cans *without washing them.* Then I put the lid on the cans and cover everything with six inches of straw, adding more when the ground freezes. I add nothing to the cans, unless the vegetables are dry at the time of storage. If their skins are dry, I sprinkle the vegetables with water as I put them in the cans, being careful not to form a pool of water at the bottom of the cans. During the winter, especially if it is dry, it is a good idea to sprinkle a little water over the vegetables once and a while. But I have found little dehydration; almost all the vegetables remain firm and fresh until the middle of March.
>
> Instead of emptying one pail at a time, I take a little from each in turn, so that the level in each can is lowered as the winter progresses. The lower the contents are, the less chance of damage from frost. If the vegetables are six to eight inches below the bin cover, with even a little straw, it will take a very severe freezing to cause frost damage because the heat from the earth below the frost line will feed into the bins.
>
> While I cannot equal or even duplicate the airtight, atmospherically controlled and refrigerated rooms that large, commercial operations boast of, my simple, inexpensive garden storage bin gives me results. All through the winter months and into early spring, long after the average gar-

dener has all but fogotten his garden, I bring in those garden-fresh, organically grown vegetables from the storage bins of my garden. Red beets, the best keeper of them all, I have enjoyed into May—weeks after the new crop was up and growing.

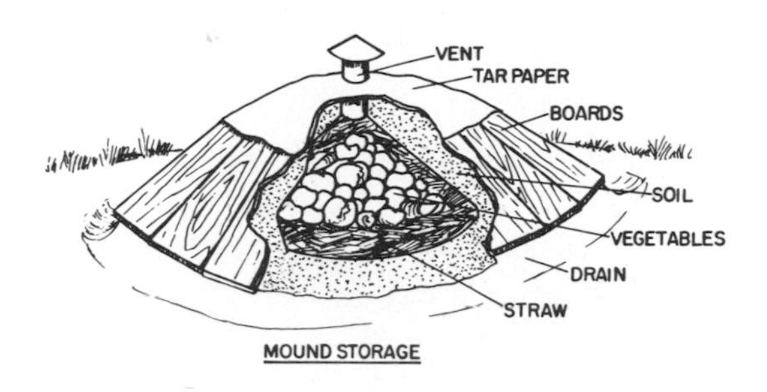

Once the mound is opened, all the food inside should be removed. Place just enough fruit or vegetables to last your family 1 or 2 weeks in each storage area.

According to Gordon Morrison, of Michigan, garbage cans or similar containers don't even have to be underground to make good storage bins. He keeps apples, potatoes, carrots and the like perfectly crisp from fall until early spring just by using watertight wooden or metal barrels, watertight butter-tubs, and candy pails. He puts them in a cool basement room, a sheltered back porch or garage—anywhere that they can safely be protected against freezing.

He provides enough ventilation by topping off the contents of the barrel with a well-moistened, but not drippy, four or five-inch head cover of something like burlap or moistened sphagnum moss, which can be rewetted when necessary. By this method he's kept many kinds of garden crops in good condition in a cement-floored basement room—next to the furnace room—where temperatures range quite often between 40 to 50° F.

Mound Storage

Other gardeners have had success storing root crops, and pears and apples right on the ground in a mound storage construction. To make a mound storage, place straw, hay, or dry leaves on the

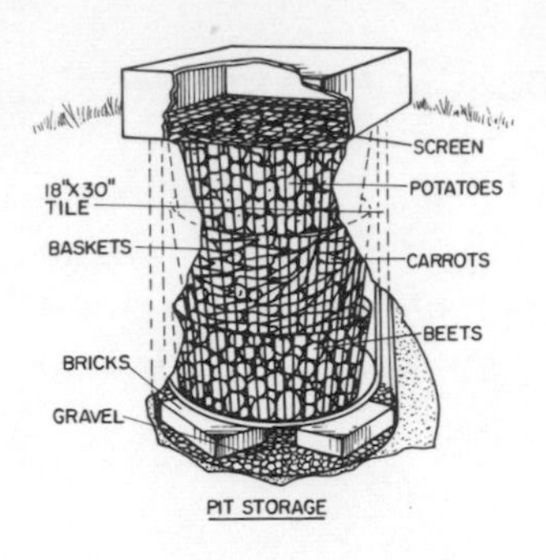

A more permanent version of Mr. Onstott's garbage pail idea is the tile pit storage. Place large (18-by-3-inch, or 24-by-24-inch)tiles in a pit on well drained soil. Positioning should be near the kitchen and shaded from the sun. After the pit is filled with boxes or baskets of food, place wire screen over the top, and cover the screen with a large rock or concrete tile.

ground, place fruit or vegetables on top of this, and cover with more mulch. Cover the mound with soil, then boarding. A ventilating pipe can be added by placing stakes or a pipe through the center of the pile. This should be capped in freezing weather to prevent cold temperatures from entering. A trench is then dug around the mound for drainage.

Once the pit is opened, all the food should be removed. It is therefore a good idea to build a number of these, each containing a small number of different vegetables. Separate the various vegetables in the mound with mulch. Do not mix fruits with vegetables, but make separate mounds for them. Mounds should be made in different places every year, because leftovers in used mounds are usually contaminated.

PICKLES AND RELISHES

Pickles and relishes have been enjoyed by millions throughout the centuries. The Chinese are said to have invented pickling—Chinese laborers who worked on the Great Wall of China carried salted vegetables to the Wall with them. In the United States, perhaps the ethnic group most famous for its pickles are the Pennsylvania Dutch. Every Pennsylvania Dutch cookbook contains dozens of recipes for both fruit and vegetable pickles, and every meal served by the Pennsylvania Dutch is supposed to contain the "seven sweets and seven sours." (There doesn't have to be seven of either; the cook picks out the best pickles and relishes to go with the meal he or she is serving.)

There are actually four different kinds of pickled products:

Brined pickles, which include sauerkraut, pickling cucumbers, and green tomatoes, go through a curing process of about three weeks. Most brined pickles are made in a low salt brine and do not require desalting before they are used. Cucumbers may also be cured in a high salt (10 percent salt) brine; these should be soaked in water before they are processed any further.

Fresh-pack or quick-process pickles are the easiest to prepare. They are usually soaked in a low salt brine for several hours or overnight, then drained and processed with boiling hot vinegar, spices, herbs or other seasonings.

Fruit pickles are usually prepared from whole fruits—pears, peaches, and watermelon rind are good choices—and simmered in spicy, sweet-sour syrup.

Relishes are mixed fruits and vegetables which are chopped, seasoned, and then cooked. Relishes may be hot and spicy or sweet and spicy. Familiar ones include picalilli, chutneys, corn relish, catsup, and the famous Pennsylvania Dutch chow chow.

INGREDIENTS

Fruits and vegetables:

Use only tender fruits or vegetables that are in prime condition. Produce should not be more than twenty-four hours old and should be refrigerated or cooled immediately after picking. This is especially important for cucumbers, which deteriorate rapidly after picking at room temperature. You may, if you wish, sort the produce for uniform size. Some canners—those who win prizes at county fairs—always sort for uniformity. By using vegetables or fruits of the same size you can be sure that all the food will cook and cure evenly.

Do not use any fruits or vegetables for pickling whole that have mold damage or are injured. Damaged ones can be cut up for relishes or, in the case of cucumbers, sliced for bread and butter pickles, with the injured or moldy part being cut out and discarded. Although proper processing does kill the spoilage organisms in moldy parts, an off-flavor will develop from the mold growth that cannot be masked by spices or herbs.

Fruits and vegetables should be washed thoroughly under running water. Scrub the food with a soft brush or with the palms of your hands. Rinse well so that all soil drains off. It is important that the food be handled gently so that it does not become bruised; this is especially important for cucumbers or for soft fruits like peaches or pears (although the latter two may be picked slightly underripe for pickling). Be sure to remove blossoms from cucumbers—the blossoms may be a source of spoilage enzymes.

Produce should never be picked for canning after a heavy rain because it will be waterlogged. Wait twelve hours after the rain to pick your fruits and vegetables.

When picking cucumbers, *cut* them from the vines, leaving a short bit of stem on the fruit. If pulled off the vines, they are likely to rot where the stem was broken from the skin.

Drain your produce on a tea towel or on a dish drain. Wipe them dry if you must, but be careful not to bruise them.

Vinegar: As practical and self-satisfying as it might be to use your own homemade vinegar in your pickles, *don't.* Vinegar should be a good grade of 40 to 60 grain strength (4 to 6 percent acetic acid). The vinegar you make on your homestead may be some of the best you've ever had, and it is fine for your salads and other cooking. But homemade vinegars vary in acidity, so you'll never know what the acidity of your vinegar is unless you test it, or have it tested. It is much safer to rely on commercial vinegar for your pickles.

Cider vinegar has a good flavor and aroma, but it is not good for white pickles such as onion or cauliflower because it may discolor them. Distilled vinegar is clear so there is no chance of discoloration. The fact that it is slightly more acidic than cider vinegar makes little difference.

Honey: Light honeys, such as clover or alfalfa, are very mild in flavor and are good to use for canning or pickling for that reason. Dark honeys are strongly flavored. If they are used in pickling, you might find that their flavor will overpower the flavors of the other ingredients. Therefore, no matter what kind of honey you are using, we suggest that you taste your syrup as you add the honey to it. If you think that the syrup is sweet enough, stop adding honey, although you may not have reached the amount suggested in the recipe. Those amounts are intended to be guidelines only, not hard and fast rules. Your best guides are your tastebuds and a knowledge of your family's likes and dislikes.

Pickling recipes from most cookbooks suggest that you boil the syrup, consisting of vinegar, spices, and sugar, for a certain amount of time. This allows the sugar to dissolve completely and also allows the spices to flavor the syrup. If you are altering a recipe that you've found in another book because you want to use honey in it rather than sugar, remember that the heating of honey tends to break down the sugars in honey and cause a change in flavor and a darkening in color. Honey can be heated to high temperatures for short periods without causing too much damage, but it will not stand sustained boiling. For this reason, never add the honey to the syrup *before* it is boiled. Instead, boil the vinegar and spices together for the stated time; then add the honey, tasting the syrup as you add it to determine sweetness. (As a general rule of thumb, if you are substituting honey for sugar, cut the amount of sweetener by one-half.) Bring the syrup to a boil and pour it over the pickles. Process as directed.

Spices: Always use whole, fresh spices or herbs. Whole spices or herbs should be tied in a bag (cheesecloth will do) or a stainless steel spice ball and removed before the pickles are packed. Never use ground herbs or spices; they tend to darken pickles. Whole spices, if left in the jar after the pickles are canned, may cause an off-flavor in the product.

If garlic is used, it should first be blanched for two minutes or removed from the jar before it is sealed.

The liquid should be tasted first before canning. Spices vary considerably in strength, and you can correct the seasoning by adding more spices. Unused spices should be kept in airtight jars in a cool place, as heat and humidity tend to sap their quality.

Water: Soft water is recommended for best-looking pickles. Iron or sulphur in hard water will darken pickles.

Salt: You should use plain salt. Iodized salt may cause darkening, and table salt sometimes causes a cloudy brine. Rock salt, dairy salt, or pickling salt is best. You can also use sea salt.

EQUIPMENT

You shouldn't need any more specialized equipment for pickling than you already have to do your other canning chores. Use unchipped enamelware or stainless steel for heating liquids you may be using in your pickling. Do not use copper, brass, galvanized, iron, or aluminum utensils. These metals react with acids and salts in the liquids and may cause undesirable color changes in the finished product.

For fermenting or brining use a crock, stone jar, plastic container, unchipped enamelware pan, or large glass jar, bowl, or casserole. Use a heavy plate or large glass lid that fits right inside the rim of the crock and a weight to hold the vegetables in the brine. Clean rocks or a glass jar filled with water can be used as a weight, or a large plastic bag filled with water can serve as both weight and cover.

If you are using recipes that specify ingredients by weight, a small household scale is a necessity. It is best to use a scale to make large quantities of sauerkraut to insure proper proportion of salt and cabbage.

CANNING PICKLED FOODS

Jars should be filled, as recommended in the recipes, leaving the necessary headspace. Pack jars firmly and uniformly, but avoid packing so tightly that the brine or syrup is prevented from filling spaces around and over the product. Wipe the rims and threads of the jar with a clean, hot cloth, and cap. Pickles are processed in a boiling-water bath. (See the instructions on boiling-water bath processing in the chapter on canning fruits and vegetables.)

Processing times in the following recipes are given for altitudes less than 1,000 feet above sea level. At altitudes above 1,000 ft., add one minute of processing time for each 1,000 feet. For example, at 2,000 ft., process for two minutes more than required time; at 3,000 ft., process for three minutes more, at 4,000 ft., process for four minutes more.

STORING PICKLES

Pickled products should be stored in cool, dark, dry places. Extreme fluctuations of temperature may cause a breakdown of texture, resulting in an inferior product. It might also cause enough expansion of the product to break the jar or the seal. Light causes products to fade and become less-appetizing in appearance. This does not mean that the product is spoiled, however.

A storage area in a basement is fine if it is cool, dry, and dark. Do not store canned products with vegetables or fruits that require high concentrations of moisture. Dampness may rust enclosures and cause spoilage.

BRINE CURING

Brine-Cured Pickles

There are many vegetables that can be cured this way: cucumbers, green tomatoes, green beans, onions, and cauliflower. This recipe is from a booklet prepared by The North Carolina Agricultural Extension Service.

1. Wash vegetables carefully; avoid bruising.
2. Weigh vegetables, put in a crock and cover with a 10 percent brine solution—made by dissolving 1 cup salt to 2 quarts water. *Brine in which a fresh egg floats is approximately 10 percent.*
3. Weight vegetables down under brine with a plate or something similar.
4. Then, or not later than the following morning, add salt at the rate of 1 cup for each 5 pounds of vegetables. This is necessary to maintain a 10 percent brine solution. *Salt should be added on top of plate to prevent its going to the bottom and forming too strong a brine there.*
5. Remove scum when it forms on top of brine. This, if left on, will destroy the acidity of the brine and result in spoilage of the product.
6. At the end of a week and for 4 or 5 succeeding weeks, add

¼ cup salt for each 5 pounds of vegetables. Add in same manner as in No. 4.

7. Fermentation resulting in bubble formation should continue about 4 weeks.
8. Vegetables may be kept in this 10 percent brine solution— no more salt added after they are cured— until made into pickles.
9. The best temperature for brining vegetables is about 80–85° F.

Vegetables may be added during the curing process if enough brine is added to cover them, *and* if salt is added in definite amounts to maintain a 10 percent brine.

Some books suggest using alum or lime for crisper pickles. It really isn't necessary to add either of these if proper procedures are followed. If you do want to add something to your pickles to crisp them, try using grape or cherry leaves during the brining process.

To desalt these pickles for further use, soak them in cold water for a few hours. You can hasten the soaking process by using large amounts of water (three to four times volume of water as volume of pickles), changing the water often, and stirring often, being careful not to bruise the pickles.

Brine-Cured Dill Pickles

Euell Gibbons, nationally-known author and expert on wild foods, wrote in his April, 1969, column in ORGANIC GARDENING AND FARMING magazine about his own adventures with pickled vegetables, using his "dill crock." This method has great appeal because it not only brines the vegetables, but flavors them as well. Gibbons writes:

> Naturally, I got started at this tasty sport with wild foods. A nearby patch of wild Jerusalem artichokes had yielded a bumper crop, and I wanted to preserve some. I used a gallon-size glass jar, getting all of these jars I wanted from a nearby school cafeteria.
>
> Packing a layer of dill on the bottom of the jar, I added several cloves of garlic, a few red tabasco peppers, then some pared and peeled artichokes, plus another layer of dill.

With room still left, I looked around for other things to add. The winter onions had great bunches of top sets, so I peeled a few and made a layer of them. Then I dug up some of the surplus onions and used the bottom sets—shaped like huge cloves of garlic—to make still another layer. I then put in a layer of cauliflower picked apart into small florets, and added some red sweet pepper cut in strips, along with a handful or so of nasturtium buds.

This was all covered with a brine made by adding three-fourths of a measure of sea salt to ten measures of water. I added some cider vinegar too, but only a quarter-cup to the whole gallon. I topped the whole thing with some more dill, set a small saucer weighted with a rock on top to keep everything below the brine, and then let it cure at room temperature.

After two weeks I decided it must be finished. The Jerusalem artichokes were superb, crisp and delicious. The winter onions, both the top and bottom sets, were the best pickled onions I ever tasted. The cauliflower florets all disappeared the first time I let my grandchildren taste them, while the nasturtium buds make better capers than capers do.

The next summer I determined to get started early and keep a huge dill crock running all season. Any size crock can be used, from one-gallon up. I use a ten-gallon one and wish it were bigger. Never try to use a set recipe for a dill crock, but rather let each one be a separate and original "creation." I plant plenty of dill, and keep planting some every few weeks so I'll always have some on hand at just the right stage.

What is good in a dill crock? Nearly any kind of firm, crisp vegetable. Green beans are perfect, and wax beans also very good. These are the only two things cooked before being added to the brine, and they should be cooked not more than about 3 minutes. And small green tomatoes are great. Nothing else so nice ever happened to a cauliflower. Just break the head up into small florets, and drop it into the dilled brine. In a week or two—the finest dilled cauliflower pickle ever tasted.

If you have winter onions, clean some sets and put them in the crock. It's tedious job, but the results are worth it. Not only do they add to the flavor of all the rest of the ingredients in the jar, but the little onions themselves are superb. If you don't have winter onions, you can sometimes buy small pickling onions on the market and use them. If not, just take ordinary onions and slice them crosswise into three or four sections. These will come apart after curing, but so what? They are simply great pickled onion rings. I've even cut off the white part of scallions and thrown them in the brine, with some success, and one late-fall dill crock was flavored with white sections of leek, which did it wonders.

To preserve these pickles, pack them in hot, sterilized jars along with some dill. Strain the brine, bring to a boil, and pour over pickles, leaving a half-inch headspace. (You can also make new brine using ½ cup salt and 4 cups of vinegar to 1 gallon of water, but the old brine is much more flavorful.) Seal. Process in boiling water for 15 minutes (start the count from the time the hot jars are placed in the actively boiling water). Remove jars and complete seals.

Spiced Vinegar

This is for use with brined pickles after they have been soaked in water for a few hours (desalted).

To 1 gallon of vinegar add a spice bag containing:

½ ounce allspice
½ ounce cloves
1 stick cinnamon
1 piece mace

Boil vinegar and spices for 15 minutes. Add

1 cup honey for sour pickles, or
2 cups honey for less acid pickles, or
4 cups honey for sweet pickles.

Heat slightly, set aside for three weeks before removing spice bag.

Sweet Cucumber Pickle

Put desalted cucumbers into sterilized jars. Bring spiced vinegar—about 1 pint vinegar per quart of pickles—to a boil. Pour over cucumbers and let them set until following day. Drain off vinegar and add honey at the rate of ¼ cup per pint of vinegar.

Bring to a boil and pour over cucumbers. On the following day again drain off vinegar and add honey in the same proportion. Seal jars. Place jars of pickles in a boiling-water bath and process 10 minutes at simmering temperature—about 180°F.

Sour Cucumber Pickle

Select *small* brined cucumbers and soak to desalt in cold water for a few hours.

To 6 quarts of cucumbers use 1 gallon of plain vinegar. Prepare spice bag as directed in spiced vinegar recipe.

Bring vinegar and spices to boil. Add the cucumbers, 1½ quarts at a time, and let them boil for 2 minutes. Don't allow them to get soft. Place them in a large stone crock or in glass jars as they are taken from the kettle. When all the cucumbers have been packed, cover them with the boiling vinegar. If a sweeter pickle is desired, add 2 cups of honey to the vinegar just before it is poured over the pickles. Place the spice bag in the jar.

Seal the jars, or cover the top of the stone jars with layers of thick paper tied tightly to exclude air. Let pickles remain in the vinegar solution for 3 weeks. Then, pack the pickles in hot, sterilized jars. Remove the spice bag from the vinegar, bring to a boil, and pour vinegar over the pickles, leaving half-inch headspace. Seal. Process for 10 minutes at simmering temperature, about 180°F.

Pickled Nasturtium Buds (False Capers)

If you pride yourself on your gourmet cooking, one recipe you can't afford to pass up is the following. As Euell Gibbons has already noted, "Nasturtium buds make better capers that capers do." Nasturtium buds should be gathered while they're still green—yellow ones are useless.

Place the nasturtium buds in a 10 percent brine—made by adding 1 cup salt to 2 quarts water—to cover. Weight them, if necessary, to hold them in the brine. Allow them to cure in brine for 24 hours.

Remove from brine, and soak in cold water for an hour. Drain the buds. Bring vinegar to a boil. Pack nasturtium buds in hot, sterilized jars, cover with boiling vinegar, leaving half-inch headspace. Seal. Process 10 minutes in a boiling-water bath, counting time from when the jar is placed in the actively boiling water.

Remove from boiling-water bath. Adjust lids, if necessary. Cool on wire racks and store in a cool, dry place. It is best to let your "capers" stand for six weeks before you use them, as they will be more flavorful. You can, however, use them immediately if you wish.

Sauerkraut

Sauerkraut has been made and enjoyed by millions for hundreds of years. For our ancestors, it was an important winter source of vitamin C and was used as a cure for scurvy on sea voyages. More recent research on sauerkraut has found that the fermentation process causes little loss in the vitamin C content of the kraut. Research has also shown that sauerkraut is very important in our diets because it has, like yogurt, a high lactic acid content, which aids digestion and helps to kill harmful bacteria in the digestive tract.

There are basically two ways to make sauerkraut: in large quantities and in small batches. We give instructions for both methods. The recipe for making small batches, suggested by Beatrice Trum Hunter, has the added advantage of being made without salt.

Saltless Sauerkraut (small batches)

Assemble a few simple items. Locate a bowl, pot or other widemouth container that will hold a gallon of liquid measure. Glass, well-glazed clay or other impermeable material is suitable. A stoneware crock and cover is ideal for the purpose. If you consider purchasing a used stoneware crock, check its inside. If the crock has been used to store surplus eggs in waterglass, it will have a permanent whitish stain. Reject such a crock for sauerkraut production.

Next, find a flat plate, slightly smaller in diameter than the inner surface of your container. If you plan to use a sloping bowl, measure the plate against the diameter near the top of the bowl. Then locate a few large, smooth stones. Plan to keep the container, plate, and stones reserved exclusively for "Project Sauerkraut."

Depending upon the amount of kraut you want to make, shred a whole or even a half of a solid head of cabbage. Place it at the bottom of the container. For each head of cabbage, pulverize a half teaspoonful each of dill, celery and caraway seeds in an electric seed grinder or with mortar and pestle. These seeds can be added whole, but the ground seeds release more flavor. If the flavor of such seeds doesn't appeal to you, omit any or all of them. If you do use them, mix with a teaspoonful of ground kelp for each head of cabbage. Sprinkle this blend on top of the shredded cabbage.

Now pour cold water over this mixture, so that the cabbage is completely covered: two quarts of water, more or less, for each cabbage. The liquid should reach no higher than two or three inches from the top of the container, to prevent overflow during fermentation.

Put the plate over the shredded cabbage, seasoning and water. Press down gently, so that the liquid flows over and submerges it. Next, weight it down with the freshly scrubbed stones. Cover the container, place it in a warm room, and let nature take its course. As a precaution against overflow, which rarely occurs with this method, place the crock in a glass pie plate to catch any drippings.

Within a few days you will begin to sniff the pleasant fermentation process. Take a peek now and then: Be sure that the plate keeps the cabbage submerged under the liquid. Skim off any scum that develops on the surface. The length of fermentation will be determined by the room's temperature. Sauerkraut may be ready in as little as seven days. Sometimes it takes a little longer. Toward the end, check daily.

When you're convinced that the product is well fermented, remove the stones and plate. Using a slotted spoon, transfer the drained sauerkraut to a bowl. Strain the remaining liquid. The flavor of both the saltless raw sauerkraut and its juice will be subtle and delicate. If you're unaccustomed to saltless food, you may find the flavor "flat." Nothing prevents you from adding a dash of sea salt. Or, if you wish, simply make the sauerkraut as described, adding sea salt instead of, or in addition to, the kelp. But remember that salt apparently is not necessary for the fermentation.

After your family eats its fill of sauerkraut, any remainder can be refrigerated. Although raw, it keeps well, due to its natural preservative, lactic acid.

Other raw vegetables can be added to ferment with the cabbage. Good additions are thinly sliced onions, carrots and turnip strips, cauliflower segments, and radish slices. Do your own experimenting.

Sauerkraut (six quarts)

If your grandmother made sauerkraut, it is likely that she used this recipe or one very close to it. Huge stoneware crocks of sauerkraut were a common sight in spring houses and cellars on farms in the past, and farm children were assigned the task of skimming bad kraut and scum off the tops of these crocks. This recipe has been adapted from one sent us by Susan Ferris of the Maine Organic Food Association.

Select 15 pounds of firm, green cabbages. (This will yield about 30 quarts of shredded cabbage.)

Let stand at room temperature for one day. Wash, quarter, remove cores. Cabbages should be dry before grating for sauerkraut. Shred or cut about the thickness of a dime.

Thoroughly mix 3 tablespoons plus 2 teaspoons salt (use pickling/canning salt) with each 10 quarts of shredded cabbage. As each batch is salted, get ready 4 to 6 quart crocks. Pack the cabbage firmly, but not tightly, into the crocks, pressing down with a wooden spoon or paddle.

Lay a clean cloth over the cabbage with a plate on top that fits just inside the crock. It is important that the cabbage is covered by the tight-fitting plate; it may spoil otherwise. Weight with a stone or a gallon jar filled with water. The weight should be heavy enough so that the liquid just reaches the bottom of the cover. To vary the weight, use heavier or lighter stones or fill or empty the jar as needed as fermentation increases.

Allow cabbage to ferment at room temperature (68 to 72° F.) for 9 to 14 days. (The lower the temperature, the slower the fermentation.) Change and wash the cloth, adjust the weight, and skim off the scum daily. Fermentation has ended when bubbles stop rising to the surface. Taste at the end of a week and can when taste suits you.

To can your kraut, use hot, scalded quart jars. Bring kraut to boil with 3 quarts water. Pack lightly into jars, filling spaces with liquid. Process in boiling-water bath for 15 minutes.

Sauerkraut (16 to 18 quarts)

This recipe is from a booklet prepared by the U.S. Department of Agriculture.

About 50 pounds of cabbage   1 pound (1½ cups salt)

Remove the outer leaves and any undesirable portions from firm, mature heads of cabbage; wash and drain. Cut into halves or quarters; remove the core. Use a shredder or sharp knife to cut the cabbage into thin shreds about the thickness of a dime.

In a large container, thoroughly mix 3 tablespoons salt with 5 pounds shredded cabbage. Let the salted cabbage stand for several minutes to wilt slightly; this allows packing without excessive breaking or bruising of the shreds.

Pack the salted cabbage firmly and evenly into a large, clean crock or jar. Using a wooden spoon or tamper, or your hands, press down firmly until the juice comes to the surface. Repeat the shredding, salting, and packing of cabbage until the crock is filled to within 3 or 4 inches of the top.

Cover cabbage with a clean, thin, white cloth (such as muslin) and tuck the edges down against the inside of the container. Cover with a plate or round paraffined board that just fits inside the container so that the cabbage is not exposed to the air. Put a weight on top of the cover so the brine comes to the cover but not over it. A glass jar or heavy-duty plastic bag filled with water makes a good weight.

The amount of water in the glass jar or plastic bag can be adjusted to give just enough pressure to keep the fermenting cabbage covered with brine.

Formation of gas bubbles indicates fermentation is taking place. A room temperature of 68° to 72° F. is best for fermenting cabbage. Fermentation is usually completed in 5 to 6 weeks.

To store the sauerkraut, heat to simmering (185° to 210° F.). Do not boil. Pack hot sauerkraut into clean, hot jars and cover with hot juice to half-inch of top of jar. Adjust jar lids. Process in boiling-water bath, 15 minutes for pints, and 20 minutes for quarts. Start to count processing time as soon as hot jars are placed into the actively boiling water.

Remove jars and complete seals if necessary. Set jars upright, several inches apart, to cool.

Sauerkraut by the Quart

Here's a simple, sure-fire way to make sauerkraut: pack it and let it ferment right in the canning jar!

Simply use quart glass jars with rubber rings and zinc lids. After washing the jars, place them upside down in a large pan of water with the rubber rings and lids, and slowly bring to a boil.

When enough cabbage is cut for several jars, pack and press the cabbage into each sterilized jar.

When the juice begins to show as it is squeezed out of the cabbage, continue to fill and press until 1 inch of space is left at the top of the jar. Then add to each quart jar 1 teaspoon of salt and ½ teaspoon of honey. Fill slowly with boiling water, allowing it to settle down into each jar. Insert a knife blade at intervals to allow air bubbles to escape. Leave about ½ inch of space at top, put the rubber ring in place and screw the zinc lid down tight.

Wipe the jars, set them in an old dish pan or a small tub in a cool corner of the garage and wait. Inspect the jars every few days. Don't be alarmed when the zinc lids begin to bulge—that shows the kraut is fermenting properly. If (and it frequently happens) the juice spews between the rubber ring and lid, wipe it off and tighten the jar lid some more. That is why the jars are set in a container outside—there's plenty of kraut odor! This will continue as long as the kraut is working. Never under any circumstances loosen the lids.

After about 6 weeks the kraut will have cured and the jars can be washed and brought inside to store with other canned foods. Don't break the seal by tightening any lid that doesn't have a fresh spew.

What You Did Wrong If Your Sauerkraut Spoiled

Spoilage of sauerkraut is indicated by undesirable color, off-odors and soft texture. If your kraut has spoiled, here's what you might have done wrong:

If your sauerkraut is soft, it may be due to insufficient salt. Try using more salt the next time. High temperatures during fermentation may also cause softness. Uneven distribution of salt may also be a cause of softness—be sure that your salt is well-mixed with the kraut next time. Air pockets caused by improper packing may also make your kraut soft. Your crock or jar may have had air spaces that caused poor fermentation; this can be remedied by packing the jar or crock tightly and being sure to weight it properly.

Pink kraut is caused by the growth of certain types of yeast on the surface of the kraut. These yeasts may grow if there is too much salt, if there is an uneven distribution of salt, or if the kraut is improperly covered or weighted during fermentation.

Rotted kraut is usually found at the surface where the cabbage has not been covered sufficiently to exclude air during fermentation.

Darkness in kraut may be caused by unwashed and improperly trimmed cabbage. It may also be caused by insufficient brine in the fermenting process. Be sure that the brine completely covers the fermenting cabbage. Exposure to air or a long storage period may also result in darkened kraut. Another cause of darkening may be high temperatures during fermentation, processing, or storage.

FRESH FRUIT AND VEGETABLE PICKLES

Sweet Gherkins

Gherkins, immature cucumbers, are often made into sweet pickles.

5 quarts (about 7 pounds) cucumbers, 1½ to 3 inches in length
½ cup salt
4 cups honey
6 cups (1½ qts.) vinegar
¾ teaspoon turmeric
2 teaspoons celery seed
2 teaspoons whole mixed pickling spice
8 1-inch pieces stick cinnamon
½ teaspoon fennel (optional)
2 teaspoons vanilla (optional)

First Day

Morning: Wash cucumbers thoroughly; scrub with vegetable brush. Drain cucumbers; place in large container and cover with boiling water.

Afternoon: Six to eight hours later, drain, cover with fresh, boiling water.

Second Day

Morning: Drain, cover with fresh, boiling water.

Afternoon: Drain, add salt, and cover with fresh, boiling water.

Third Day

Morning: Drain. Prick cucumbers in several places with table fork. Add turmeric and spices (including fennel, if you are using it) to 3 cups of vinegar and bring to a boil. Add 1½ cups of honey. Poor over cucumbers (they will be partially covered at this point).

Afternoon: Drain syrup into pan; add 2 cups of the vinegar to syrup, bring to boil. Add 1 cup of honey. Poor syrup over pickles.

Fourth Day

Morning: Drain syrup into pan; add ½ cup of the vinegar to syrup. Heat to boiling. Add 1 cup honey and pour over the pickles.

Afternoon: Drain syrup into pan. Add remaining ½ cup honey (plus vanilla if you are using it) to the syrup. Bring to quick boil. Pack pickles into clean, hot pint jars. Cover with boiling syrup, leaving a half-inch headspace. Seal.

Process for 5 minutes in boiling water, counting processing time as soon as the hot jars are placed in the boiling water. Remove jars and complete seals if necessary. Set jars upright, several inches apart, on a rack to cool.

Fresh-pack Dill Pickles (Unsweetened)

- 17 – 18 pounds cucumbers, 3 to 5 inches in length (pack 7 to 10 per quart jar)
- 2 gallons 5 percent brine (¾ cup salt per gallon water)
- 6 cups (1½ quarts) vinegar
- ¾ cup salt
- 9 cups water
- 2 tablespoons whole mixed pickling spice
- 2 teaspoons whole mustard seed per quart jar
- 1 or 2 cloves garlic per quart jar (optional)
- 3 heads fresh dill per quart jar (you may substitute dried dill) or 1 tablespoon dill seed per quart jar

Wash cucumbers thoroughly, scrub with vegetable brush, and drain. Cover with brine. Let set overnight. Drain.

Combine vinegar, salt, water, and mixed pickling spices tied in a clean, white cloth bag or stainless steel spice ball. Heat to boiling. Pack cucumbers in hot, sterilized quart jars. Add mustard seed, garlic, dill plant or seed to each jar. Cover with boiling liquid to within a half-inch of top of jar. Seal.

Process in boiling water for 20 minutes, starting to count the processing time as soon as the hot jars are placed in the actively boiling water. Remove the jars and complete seals if necessary. Set jars upright, several inches apart, on a wire rack to cool.

Crosscut Pickle Slices

4 quarts cucumbers, medium-size, sliced (about 6 pounds)
1½ cups onions, sliced
2 large garlic cloves
⅓ cup salt
2 quarts ice cubes or chips
2 cups honey
1½ teaspoons turmeric
1½ teaspoons celery seeds
2 tablespoons mustard seeds
3 cups white vinegar

Wash cucumbers thoroughly and scrub with a vegetable brush. Drain on rack. Slice unpeeled cucumbers into ⅛ to ¼-inch slices; discard ends. Add onions and garlic.

Add salt and mix thoroughly, cover with crushed ice or ice cubes and let stand three hours. Drain thoroughly and remove garlic.

Combine honey, spices (in a spice bag), and vinegar. Heat just until boiling. Add drained cucumber and onion slices and simmer for 5 minutes.

Do Not Boil.

Pack hot pickles loosely in clean, hot pint jars, leaving a half-inch head space. Seal. Process in boiling-water bath for 5 minutes, starting to count processing time as soon as the water in the canner returns to boiling. Remove jars and complete seals. Set jars upright on wire rack a few inches apart to cool.

Bread and Butter Pickles

30 cucumbers, medium
10 onions, medium
4 tablespoons salt

Slice cucumbers and onions and sprinkle with salt. Let stand 1 hour. Drain in cheesecloth bag. Make a spiced vinegar using the following ingredients:

5 cups vinegar
2 teaspoons celery seed
2 teaspoons ground ginger
2 cups honey
1 teaspoon turmeric
2 teaspoons white mustard seed

Let spiced vinegar come to a boil, add cucumbers and onions and bring to boiling point. Simmer 10 minutes. Seal in sterilized jars. Process 10 minutes at simmering temperature—about 180° F.

Old-fashioned Cucumber Chunks

1 gallon cucumbers, cut into 1-inch pieces
1½ cups salt
9 cups vinegar
2 cups honey
2 tablespoons mixed pickling spices

Wash, dry, and cut cucumbers into 1-inch pieces before measuring. Put in crock or large container. Dissolve salt in 1 gallon water. Pour over cucumbers. Cover with plate and weight so that the cucumbers remain submerged in the brine. Let stand 36 hours.

Drain. Pour 4 cups vinegar over the cucumbers; add enough water to cover. Simmer 10 minutes. Drain and discard liquid. Add spices (tied in bag) to 3 cups of water and 5 cups of vinegar. Simmer 10 minutes. Add 1 cup honey. Pour over cucumbers. Let stand 24 hours.

Drain syrup into kettle. Add remaining honey. Heat to boiling. Pour over cucumbers. Let stand 24 hours.

Pack pickles into hot, sterilized jar. Heat syrup to boiling and pour over pickles, leaving a half-inch headspace. If there is not enough liquid to cover pickles, add more vinegar. Seal.

Process in boiling-water bath for 10 minutes, starting to count time from the time the jars are placed in actively boiling water.

Cucumber Oil Pickles

100 cucumbers
3 onions
2 cups salt
2 cups honey
2 tablespoons peppercorns
4 tablespoons mustard seed
4 tablespoons celery seeds
4 cups vinegar
1 cup olive oil

Wash, dry, and thinly slice unpeeled cucumbers and peeled onions. Dissolve salt in 1 gallon cold water. Add cucumbers and onions. Let stand 12 to 18 hours. Drain. (Taste the cucumbers. If they are too salty, rinse well in cold water.)

Place spices (in spice bag), 1 cup water, and vinegar in pot. Boil 1 minute. Add honey, cucumbers, onions, and oil. Simmer until cucumbers change color. Then bring to boiling. Pack, boiling hot, into hot, sterilized jars. Seal. Process for 10 minutes in boiling-water bath; begin counting time when jars are placed in boiling water.

Dilled Green Beans

4 pounds whole green beans (about 4 quarts)
5 cups vinegar (1¼ quarts)
5 cups water
½ cup salt
¼ teaspoon crushed red pepper per pint jar
½ teaspoon whole mustard seed per pint jar
½ teaspoon dill seed per pint jar
1 clove garlic per pint jar

Wash beans thoroughly; drain and cut into lengths to fill pint jars. Pack beans into clean, hot jars; add pepper, mustard seed, dill seed, and garlic.

Combine vinegar, water and salt; heat to boiling. Pour boiling liquid over the beans, filling jars but leaving a half-inch head-space. Seal. Process in boiling water for 5 minutes. Remove jars from canner and complete seals. Set jars upright, several inches apart on wire racks to cool.

Artichoke Pickles

1 peck (8 quarts) artichokes
2 cups salt
4 tablespoons turmeric
1 gallon vinegar
1 box mixed pickling spices tied in a spice bag
2 tablespoons turmeric
6 cups honey
medium-sized pod red peppers
onions

Wash and cut artichokes. Pack in a large crock or enamel pot. Cover with vinegar. Add 2 cups salt and 4 tablespoons turmeric. Soak for 24 hours.

In the meantime, make spiced vinegar by combining in a pot 1 gallon vinegar, 2 tablespoons turmeric, and spice bag with pickling spices. Boil the mixture for 20 minutes. Remove the spice bag and add 6 cups honey. And bring the mixture to a boil.

Drain the artichokes. Pack in jars, covering with the boiling spiced vinegar. Allow a half-inch headspace. To taste to each jar add 1 medium pod red pepper and onions. Process the jars for 10 minutes in boiling-water bath.

Onion Pickle

Select small silver-skin onions and sort for size. Use onions that are a half-inch in diameter. Remove the skins until the smooth surface is reached.

Place in a large jar or crock and pour a strong brine over the onions—1 to 1½ pounds of salt per gallon of water. Let the onions stand in the brine for 24 hours.

The following day, make another strong brine, bring to boil, drop the onions in the brine, and cook for 5 minutes. Remove and drop onions in cold water for 1 hour. Drain, place in large jars, and pour boiling spiced vinegar (see previous recipe) over them. Cover with saucer or small plate and weight to keep onions submerged in brine.

Let the onions stand in a cool, dark place for about 6 weeks. Pack the onions in hot, sterilized jars. Pour fresh, boiling spiced vinegar over them, allowing a half-inch headspace in the jar. Garnish with a sprig of mace and a small hot pepper, if desired. Process 10 minutes in boiling-water bath.

Carrot Pickle

3 pounds carrots
1 pint distilled vinegar
½ tablespoon whole cloves
½ tablespoon allspice
½ tablespoon mace
1 cup honey
½ stick cinnamon

Pare carrots and cut in strips that are the desired size and length. Boil in water until just heated through. Pack hot carrots lengthwise in hot, sterilized jars. Make syrup of vinegar and spices, (in spice bag) bring to boil and simmer for 5 minutes. Add honey, bring to boil, and pour over carrots. Seal and process 10 minutes in boiling-water bath, counting time when the jars are placed in the actively boiling water.

Okra Pickle

- 3½ pounds small okra pods
- 1 pint distilled vinegar, colorless
- 1 quart water
- 2 teaspoons dill seed
- ⅓ cup salt
- 3 small hot peppers, if desired, for each jar
- 1 clove garlic for each jar

Wash okra, and pack firmly in hot sterilized jars. In each jar put a clove of garlic and hot peppers if you wish. Make a brine with the water, vinegar, dill seed and salt. Boil. Pour boiling brine over okra, leaving a half-inch headspace at top. Seal.

Process for 10 minutes in boiling-water bath. Let ripen several weeks before using.

Beet Pickle

- 1 gallon small beets
- 1 cup honey
- 1 long stick cinnamon
- 1 tablespoon whole allspice
- 1 quart vinegar

Cook beets with roots and about 2 inches of stem left on in water to cover. Cook until tender; dip beets into cold water and slip off skins. Put beets in large preserving kettle. Combine other ingredients, pour over beets and simmer 15 minutes. Pack hot into sterilized jars. Cover beets with boiling syrup and seal.

Process 10 minutes in boiling-water bath.

Sweet and Sour Cabbage

- 4 quarts finely shredded red cabbage
- 4 tart apples, diced
- 1½ quarts cider vinegar
- 1 cup honey
- 1 or 2 cups water (sufficient amount for juice)
- 4 teaspoons sea salt
- ½ teaspoon peppercorns
- 2 teaspoons caraway seed
- ½ teaspoon mace (optional)
- ½ teaspoon whole allspice
- ¼ teaspoon cinnamon log pieces

Simmer in a large canner for 20 to 25 minutes. Remove spices. Pack into hot pint jars to within 1 inch of the top. Process in a boiling-water bath for 15 minutes.

This cabbage will be much too stout for most people to eat from the jar, but is delectable when the juice is drained off and it is simmered in a small amount of water.

Dutch Spiced Red Cabbage

- 2 heads red cabbage
- ½ cup salt
- 1 gallon vinegar
- ½ cup honey
- 1 teaspoon celery seed
- 1 teaspoon pepper
- 1 teaspoon each mace, allspice, cinnamon

Shred the cabbage, sprinkle with the salt, let stand 24 hours. Press moisture out, stand in sun for 3 hours. Boil the vinegar for 8 minutes with ½ cup water and the spices. Add honey. While hot pour over the cabbage. Keep in large bowl or earthen jar or can, as for sauerkraut.

Sweet Pickled Tomatoes

Slice 1 gallon green tomatoes. Put them in a large crock or glass container. Pour over them enough water to cover. Sprinkle ¼ inch of salt on top of the water. Let stand for 24 hours.

Drain. Put the tomatoes into a large kettle. Add 2 cups honey, and enough vinegar to cover. Bring to boil, take off the heat and pour into hot, sterilized jars. Leave a half-inch headspace at top. Seal.

Process in boiling-water bath for 10 minutes. Remove from heat, complete seals if necessary. Allow to cool several inches apart on wire rack.

Watermelon Pickles

- 3 quarts watermelon rind (about 6 pounds, unpared, or ½ large melon)
- ¾ cup salt
- 3 quarts water
- 2 quarts ice cubes
- 4½ cups honey
- 3 cups white vinegar
- 3 cups water
- 1 tablespoon whole cloves
- 1 thinly sliced lemon, with
- 6 1-inch pieces stick cinnamon

Pare rind and pink edges from the watermelon. Cut into 1-inch squares or fancy shapes as desired. Cover with brine made by mixing the salt with 3 quarts cold water. Add ice cubes. Let stand 4 or 5 hours.

Drain. Rinse in cold water. Cover with cold water and cook until fork tender, about 10 minutes. Do not overcook.

Combine vinegar, water, and spices (tied in a spice bag). Boil 5 minutes, add honey, and pour over watermelon (do not remove spices). Add lemon slices. Let stand overnight.

Heat watermelon in vinegar syrup until boiling and simmer until watermelon is translucent (about 10 minutes). Pack hot pickles loosely into clean, hot pint jars. To each jar add 1 piece of stick cinnamon from spice bag. Cover with boiling syrup, leaving a half-inch headspace. Adjust jar lids.

Process in boiling water for 5 minutes. Remove jars and complete seal if necessary. Set jars upright, several inches apart, on a wire rack to cool.

Honey may be reduced if a less sweet pickle is desired.

Collect watermelon rind in plastic bags in the refrigerator until you have enough for one recipe.

Pickled Peaches

1½ quarts honey
2 quarts vinegar
7 2-inch pieces stick cinnamon
2 tablespoons whole cloves
16 pounds small or medium-sized peaches

Place spices in a spice bag. Combine vinegar and spices in a large kettle. Bring to a boil, cover, and let simmer about 30 minutes.

Wash the peaches and remove skins. (Dipping the fruit in boiling water for 1 minute, then quickly in cold water makes peeling easier.) To prevent pared peaches from darkening during preparation, immediately put them into cold water containing 2 tablespoons salt and 2 tablespoons vinegar per gallon. Drain just before using.

Add honey to the syrup and bring to a boil. Add peaches, enough for 2 or 3 quarts at a time, and simmer for 5 minutes. Pack hot peaches into clean, hot jars. Continue heating in syrup and packing peaches. Add 1 piece stick cinnamon and 2 to 3 whole cloves (if desired) to each jar. Cover peaches with boiling syrup, leaving a half-inch headspace. Seal.

Process in boiling water for 20 minutes. Remove jars and complete seals if necessary.

Pear Pickle

8 pounds pears
10 2-inch pieces stick cinnamon
2 tablespoons whole cloves
2 tablespoons whole allspice
2 pounds honey
1½ quarts honey

For seckel pears: Wash the pears and remove the blossom ends only. Boil the pears for 10 minutes in enough water to cover. Drain. Prick the skins. Put spices in bag, boil with vinegar for 5 minutes. Add the honey and bring to boil. Add pears, simmer for 10 minutes or until the pears are tender. Do not overcook. Let stand overnight.

In the morning, remove the spice bag. Drain syrup from the pears and heat syrup to boiling. Pack.pears in clean, hot, sterilized jars. Pour hot syrup over the pears, filling jars to top. Seal. process 10 minutes in boiling water bath.

For kieffer pears: Use kieffer pears, and reduce vinegar to 5 cups. Wash the pears, pare, cut in halves or quarters, remove hard centers and cores. Boil 10 minutes in enough water to cover. Drain. Proceed as for seckel pears.

Sweet Fruit Pickles

2 cups mild flavored honey	½ lemon sliced, or 3 calamondins or kumquats, cut in thick slices and seeded
1 cup cider vinegar	3 inches stick cinnamon
1 cup water	12 whole cloves
½ piece ginger root fruit	

Combine honey, vinegar, and citrus fruit and spices. Heat to boiling and simmer gently about 5 minutes. Have ready 4 to 6 cups of the quartered pears, peach halves or pineapple chunks. Add to spiced solution. Cook until just tender. Pack fruit in hot jars, cover with the boiling syrup and seal.

Process for ten minutes in boiling-water bath. Remove jars and complete seals.

Spiced Crabapples

Wash 12 pounds of crabapples well (use larger ones). Be sure to remove any blossom ends. Make a syrup by heating together:

1 quart vinegar	1 tablespoon cloves	in spice bag
2 cups honey	1 teaspoon allspice	
1 stick cinnamon	1 teaspoon mace	

When this syrup is cool, add the crabapples and heat slowly so as not to burst the fruit. Sometimes it is well to prick each apple to avoid bursting. Bring to a boil. Allow to cool overnight. Remove spice bag. Heat to boiling point. Pack crabapples into hot, sterilized jars, fill to within a half inch from the top with syrup. Seal.

Process 10 minutes in boiling-water bath.

Spiced Sweet Apples

7 pounds sweet apples, quartered and cored
4 cups honey
1 quart vinegar
2 cups water
1 ounce allspice
1 ounce cinnamon stick
cloves (about 2 for each quarter)

Put vinegar, water, and spices in pot. Bring to boil. Add honey and apples, bring to boil, and simmer gently until the fruit is tender. Place the apple quarters in hot, sterilized jars, bring syrup to boil, and pour over the apples, leaving a half-inch headspace. Seal. Process 10 minutes in boiling-water bath. Remove from heat. Adjust lids and set on wire rack to cool.

Minted Sweet Apples

Use the above recipe, but substitute 1 cup of mint tea for the spices. Make the tea by simmering 1 cup of fresh green mint in a pint of water. Strain before adding to the syrup. Reduce water to 1 cup.

Lemon Pickles

Use only organically-grown, unsprayed lemons for this recipe, as for any recipe that calls for fruit rind (marmalades, relishes, *etc.*)

12 large lemons
½ cup salt
8 garlic cloves, peeled
1 tablespoon mace
1 tablespoon allspice
1 tablespoon nutmeg, grated
1 teaspoon red pepper
4 tablespoons dry mustard
½ gallon vinegar

Wash and dry the lemons. Cut each lengthwise into 8 sections. Put these lemon sections into a pan with salt, garlic, and spices (tied in a spice bag). Add vinegar and bring to a boil. Simmer 30 minutes. Pour into large stone jar or crock. Stir daily for 1 month.

At the end of a month, drain the liquid into a pot. Place the lemons in hot, sterilized glass jars. Bring the liquid to a boil, remove spices, and pour over the lemons. Seal.

Process for 10 minutes in boiling-water bath. Remove from

canner, adjust lids, allow to dry on a wire rack. Store in a cool, dry place.

RELISHES AND CHUTNEYS

Chow Chow

- 2 small heads cauliflower, cut into small flowers
- 1 large bunch celery, cut into slices
- 2 pounds onions, sliced or 2 pounds pearl onions
- 2½ quarts fresh lima beans
- 1 quart cut green peppers
- 2 quarts yellow string beans, cut into small pieces
- 2 dozen ears corn, kernels removed
- 2 quarts carrots, chunked
- 2 – 3 quarts small cucumbers (may be brined)
- 2 quarts kidney beans
- salt
- 2 quarts cider vinegar
- 5 cups honey (less, if desired)
- 1 tablespoon mustard seed
- 1 tablespoon peppercorn
- 1 tablespoon whole cloves
- cinnamon stick

Sprinkle vegetables with salt and let stand 24 hours. Drain. Make a syrup from the rest of the ingredients. Bring syrup to boil, add vegetables, and simmer until the vegetables are soft. Pack in hot, sterilized jars and seal.

Process for 10 minutes in boiling-water bath. Remove jars, complete seals. Dry upright on wire rack, separating jars so that air can circulate between them.

Cherry Relish

- 2 cups pitted cherries
- 1 cup seedless raisins
- 1 teaspoon cinnamon
- ¼ teaspoon cloves
- ½ cup honey
- ½ cup vinegar
- 1 cup pecans

Combine all ingredients but nuts in pot. Cook slowly 1 hour. Add 1 cup broken pecan nut meats and cook 3 minutes longer. Pour into hot, sterilized jars and seal.

Process in boiling-water bath for 10 minutes. Remove from canner and complete seal if necessary.

Honey Chutney

2 quarts sour apples
2 green peppers
⅓ cup onions
¾ pound seedless raisins
½ tablespoon salt
1 cup honey
juice of 2 lemons and the grated rind of one
1½ cups vinegar
¾ cup tart fruit juice
¾ tablespoon ginger
¼ teaspoon cayenne pepper

Wash and chop fruit and vegetables. Add all other ingredients and simmer until thick. Pour into hot, sterilized jars and seal.

Process in boiling-water bath for 10 minutes. Remove from canner and complete seal if necessary.

Horseradish Relish

1 cup grated horseradish
½ cup white vinegar
¼ teaspoon salt

Wash horseradish roots thoroughly and remove the brown, outer skin. (A vegetable peeler is useful in removal of outer skin.) The roots may be grated, or cut into small cubes and put through a food chopper or a blender.

Combine ingredients. Pack into clean jars. Seal tightly. Store in refrigerator.

Corn Relish

2 quarts whole kernel corn (Use 16 to 20 medium-size ears of fresh corn or 6 10 oz. packages of frozen corn)
1 pint sweet red peppers, diced
1 pint green peppers, diced
1 quart (1 large bunch) celery, chopped
8 to 10 small onions, chopped or sliced
¾ cup honey
1 quart vinegar
2 tablespoons salt
2 teaspoons celery seed
2 tablespoons powdered mustard
1 teaspoon turmeric

For fresh corn: Remove husks and silks. Cook ears of corn in boiling water for 5 minutes; remove and plunge into cold water. Drain and cut corn from cob. Do not scrape cob.

For frozen corn: Defrost overnight in refrigerator or for 2 to 3 hours at room temperature. Place containers in front of a fan to hasten defrosting.

Combine green peppers, red peppers, celery, onions, honey, vinegar, salt, and celery seed. Cover pan until mixture starts to boil, then simmer uncovered for 5 minutes, stirring occasionally. Mix dry mustard and turmeric and blend with liquid from above mixture; add, with corn, to mixture. Heat to boiling and simmer for 5 minutes, stirring occasionally.

Bring to boil. Pack loosely while boiling into clean, hot pint jars. Allow a half-inch headspace. Seal.

Process in boiling-water bath for 15 minutes. Remove jars and complete seals if necessary.

Pepper Relish

- 1 pint coarsely ground onion
- 2 cups coarsely ground sweet green pepper
- 2 cups coarsely ground sweet red pepper
- 1½ cups honey
- 1 quart vinegar
- 2 teaspoons salt

Drain pepper and onions, then combine all ingredients and bring slowly to a boil. Simmer until slightly thickened, about 25 minutes. Pour into clean, hot sterilized jars. Fill jars, leaving a half-inch headspace. Seal.

Process 10 minutes in boiling-water bath. Remove jars from canner, check and complete seal. Set upright on wire rack to cool.

Pepper-onion Relish

- 1 quart onions (6 to 8 large), finely chopped
- 1 pint sweet red peppers (4 or 5 medium), finely chopped
- 1 pint green peppers (4 or 5 medium), finely chopped
- ½ cup honey
- 1 quart vinegar
- 4 teaspoons salt

Combine all ingredients and bring to a boil. Cook until slightly thickened (about 45 minutes), stirring occasionally. Pack the boiling hot relish into clean, hot jars; fill to top of jar. Seal tightly. Store in refrigerator.

If extended storage without refrigeration is desired, this product should be processed in a boiling-water bath. Pack the boiling hot relish into clean, hot jars to a half inch of top of jar. Adjust jar lids. Process in boiling water for 5 minutes.

Remove jars and complete seals if necessary. Set jars upright, several inches apart, on a wire rack to cool.

Tomato-Apple Chutney

- 3 quarts chopped tomatoes (18 to 20 medium-size)
- 3 quarts chopped apples (12 to 15 medium-size)
- 1 cup chopped green pepper
- 3 cups chopped onion
- 2 cups seedless raisins
- 4 teaspoons salt
- 4 cups vinegar
- ⅓ cup whole mixed pickle spices

Combine tomatoes, apples, green pepper, onion, raisins, salt, and vinegar. Put spices loosely in a spice bag and add to tomato mixture. Bring to a boil, and simmer 1 hour, stirring frequently. Remove spices. Pack chutney into clean, hot, sterilized jars. Seal and process 10 minutes at simmering temperature—about 180° F.

Tomato-Pear Chutney

- 2½ cups tomatoes, quartered, fresh or canned
- 2½ cups pears, diced, fresh or canned
- ½ cup seedless white raisins
- ½ cup green pepper, chopped (1 medium)
- ½ cup onions, chopped, (1 or 2 medium)
- ½ cup white vinegar
- 1 teaspoon salt
- ½ teaspoon ground ginger
- ½ teaspoon powdered dry mustard
- ⅛ teaspoon cayenne pepper
- ¼ cup canned pimiento

When fresh tomatoes or pears are used, remove skins; when canned pears are used, include syrup.

Combine all ingredients except pimiento. Bring to boil. Cook slowly until thickened (about 45 minutes), stirring occasionally. Add pimiento, bring to boil, and cook 3 minutes longer.

For refrigeration storage: Pack the boiling hot chutney into clean, hot jars, filling to the top. Seal tightly.

For canning: Pack the boiling hot chutney into clean, hot jars, leaving a half-inch headspace. Seal. Process in boiling-water bath for 5 minutes. Remove jars, complete seals if necessary. Set jars upright on rack to cool.

If a slightly sweeter chutney is desired, you may add honey to taste just before canning. Stir in honey, tasting until desired sweetness is reached, just before pouring into jars. The cayenne pepper may be reduced or eliminated if a less spicy chutney is desired. The best test, again, is a taste test.

Piccalilli

- 1 quart green tomatoes, about 16 medium tomatoes, chopped
- 1 cup sweet red peppers, chopped (2 to 3 medium peppers)
- 1 cup green peppers, chopped (2 to 3 medium)
- 1½ cups onions, chopped (2 to 3 large)
- 5 cups cabbage, chopped (about 2 pounds)
- ⅓ cup salt
- 3 cups vinegar
- 1 cup honey
- 2 tablespoons whole mixed pickling spice

Combine vegetables, mix with salt, let stand overnight. Drain and press in a cheesecloth bag to remove all liquid possible.

Combine vinegar with the spices tied in a spice bag. Bring mixture to boil. Add honey. Add vegetables, bring to boil, and simmer about 30 minutes, or until there is just enough liquid to moisten vegetables. Remove spice bag. Pack hot relish into clean, hot pint jars, leaving a half-inch headspace. Seal.

Process in boiling-water bath for 5 minutes, beginning to count the processing time as soon as the water in the canner returns to boiling. Remove jars and complete seals if necessary. Set jars upright on wire rack to cool.

Chili Sauce

- 4 quarts peeled and chopped tomatoes (24 to 28 medium size)
- 2 cups chopped sweet red pepper
- 2 cups chopped onion
- 1 hot pepper, chopped
- 2 tablespoons celery seed
- 1 tablespoon mustard seed
- 1 bay leaf
- 1 teaspoon whole cloves
- 1 teaspoon ground ginger
- 1 teaspoon ground nutmeg
- 2 3-inch pieces stick cinnamon
- 3 cups vinegar
- 2 tablespoons salt
- ½ cup honey

Combine the tomatoes, sweet pepper, onion, and hot pepper. Put the celery seed, mustard seed, bay leaf, cloves, ginger, nutmeg and cinnamon loosely in a thin, white cloth; tie top tightly; add to tomato mixture and boil until one-half original volume. Stir frequently to prevent sticking. Add the honey, vinegar and salt. Boil rapidly, stirring constantly; simmer about 5 minutes. Pack into clean, hot, sterilized jars. Seal.

Process 10 minutes in boiling-water bath. Remove jars from canner and complete seal. Cool upright on wire rack.

Catsup

- 2½ quarts sliced tomatoes (15 to 17 medium-sized)
- ¾ cup chopped onion
- 3-inch piece stick cinnamon
- 1 large garlic clove, chopped
- 1 teaspoon whole cloves
- 1 cup vinegar
- 1¼ teaspoons salt
- 1 teaspoon paprika
- Dash cayenne pepper

Simmer together tomatoes and onion for about 20 to 30 minutes; press through a sieve. Put the cinnamon, garlic, and cloves loosely in a clean, thin, white cloth; tie top tightly; add to vinegar and simmer 30 minutes. Remove spices. Boil sieved tomatoes rapidly until one-half original volume. Stir frequently to prevent sticking. Add spiced vinegar, salt, paprika, and cayenne pepper to tomato mixture. Boil rapidly, stirring constantly, about 10 minutes or until slightly thickened. Pour into clean, hot, sterilized jars. Seal.

Process for 10 minutes in boiling-water bath. Remove from canner. Complete seals if necessary. Cool on wire rack.

If you want a somewhat sweeter catsup, you can add honey

just before you seal. Your taste is your best guide here, so add honey, stirring it in until the taste pleases you.

IF YOUR PICKLES FAIL

If your pickles, for some reason, don't turn out the way you'd like them too, learn from your mistakes and don't do the same thing wrong next year. Here are some common causes of pickle failure and how you can correct them:

If your pickles are shriveled, you may have used too strong a vinegar or salt solution at the start of the pickling process. In making very sweet or very sour pickles, it is best to start with a dilute solution and increase gradually to the desired strength. Shriveling may also be caused by overcooking or overprocessing.

Hollowness in cucumber pickles usually results from one or several of the following: poorly developed cucumbers, holding the cucumbers too long before pickling, too rapid fermentation, or too strong or too weak a brine during fermentation.

Soft or slippery pickles are spoiled pickles. Do not use them. This condition is generally a result of microbial action which caused the spoilage, and it is irreversible. Proper processing should halt microbial activity, but if it results, here are some things you might have done wrong: used too little salt or acid, failed to cover your cucumbers with brine during fermentation, allowed scum to scatter through the brine during fermentation, processed the pickles for too short a time, did not seal the jar air tight, or used moldy garlic or spices. Also, if you failed to remove the blossoms from the cucumbers before fermentation, they may have contained fungi or yeasts responsible for the softening action.

Dark pickles are not spoiled pickles; however, if you are one of those people who prides himself on the looks of his home-canned products, darkening can be annoying. Darkening may result from the use of ground spices, too much spice, iodized salt, overcooking, iron in the water you used, or the use of iron utensils.

JAMS JELLIES AND FRUIT BUTTERS

What's a book on food preservation without a chapter on jelly-making? You may ask, and rightly so. In fact, if you've looked at most books on food preservation, you've probably found that the chapter on making jellies and jams is the largest section in the entire book (except, perhaps, for the pickling section). This is because jelly-making offers a chance to be truly creative. As long as you follow some basic rules, you can explore some very interesting flavor combinations of fresh fruits, dried fruits, and nuts. Perhaps you remember the pride your mother had in a special jelly, or the smile on your grandmother's face as she served slabs of homemade bread spread thick with butter and her own peach jam. Jellied fruit products have been a mark of pride for farm women for years. Even today, jams and jellies are an important item of show in farm shows and county fairs throughout rural areas of the country.

Making jellied fruit products—jams, jellies, conserves, preserves, marmalades, and fruit butters—is an excellent way of using up fruits and berries that can't be used for other types of canning. When you size your fruit for canning you'll most likely have some fruit that is too small or too large to be canned. Some pieces may be rejected because they have imperfections, like soft spots or rotten portions. Other pieces may be too ripe for canning. All these fruits can be used for jelly-making, and all of them can make a definite contribution to the finished product. Ripe fruit adds to the flavor of the product and should always be used when you are using commercial pectins. And, of course, size or blemishes won't matter because all the fruit is going to be cut up and peeled anyway.

THE COOKED-DOWN METHOD

There are basically two ways of making jellies, jams, preserves, conserves and marmalades (fruit butters are made differently). The first way is the traditional, cooked-down method. When using this method, select your fruit carefully. A batch of fruit should contain one-fourth just-ripe fruit and three quarters fully ripe fruit for best pectin and flavor content. It is best to make jellies and jams by the cooked-down method only with fruits that have pectin content. These fruits include apples, blackberries, grapes, quinces, and raspberries. The fruit must also contain a sufficient amount of acid to achieve a gel. If the fruit you're working with is low in acid, recipes will call for the addition of lemon juice or citric acid to make up for the lack of acid. Cooked-down jellies also require sugar to gel. Generally, home-made jellies or jams contain about 40 percent sugar when they are finished; more sugar tends to make the finished product tough and will tend to mask the flavor of the fruit. Besides helping the gelling process, sugar also acts as a preservative.

USING COMMERCIAL PECTIN

The second way you can make jellies or jams is by using a commercially made pectin, which is available in liquid and powder form in most food stores. Jelly made with commercial pectin requires less cooking time than cooked-down jelly, and is easier to use. Jellies made with commercial pectin require less fruit than the cooked-down type to make the same amount. This pectin will also achieve a gel with fruits that are low in pectin, such as strawberries and apricots. However, more sugar is needed to achieve a gel if commercial pectins are used. Some manufacturers, in fact, suggest that you use 60 percent sugar to 40 percent fruit juice in jellies made with their product. (By comparison, commercially manufactured jellies must contain at least 45 percent fruit and 55 percent sugar.)

TO USE OR NOT TO USE SUGAR

Most cookbooks neglect to underscore the amount of sugar you need to make jellies. Don't be too surprised if you find recipes calling for, say, one and one-half quarts of strawberries and five cups of sugar. That same recipe may yield five half-pints of preserves, which means that there is over a cup of sugar in every half-pint. That's an awful lot of sugar!

Because of its sweetness, sugar was a highly prized commodity for centuries before its preservative powers were discovered. It was always included with trade goods for the Indians, for example, because Indians prized it so highly. When it was discovered that sugar did a good job of preserving foods, it became even more valuable on farms and homesteads. During World War II, when sugar supplies were low and what little there was had to be rationed, government researchers began experimenting with alternatives to sugar. Books on preserving written during the war tell how to substitute honey or corn syrup (another refined product) for some of the sugar in jelly and jam recipes.

It is becoming more apparent, however, that serious effort must be made to find a real, workable alternative for sugar for many uses in the home. Evidence is mounting that points to sugar as a real danger to health. For years experts have known the role sugar plays in tooth decay, diabetes, and obesity. Now there are indications that sugar may be responsible, in part, for coronary disease, and it has even been suggested that sugar may be a cause of cancer in some people. There is, in short, growing medical evidence that we would be wise to avoid eating any sugar at all*

Even though medical experts and nutritionists don't know precisely to what extent sugar does our bodies harm, they do know that sugar does us no good. Sugar, even the so-called "raw" sugar, has no significant nutritional value. The complex sugar refining process destroys almost all of the few original vitamins and minerals found in the raw sugar cane and sugar beet. When you eat sugar, you're consuming pure carbohydrate.

MAKING JAMS AND JELLIES WITH HONEY

But what can be substituted for sugar in jams and jellies? We recommend honey. Honey, unlike sugar, is subjected to a minimal

*John Yudkin, *Sweet and Dangerous* (New York: Peter H. Wyden, Inc., 1972)

amount of refining and processing and contains its original vitamins, enzymes, and minerals. In its unheated, unfiltered form, it also contains pollen, a food rich in vitamins, minerals, and proteins. What's more, honey can be produced on the homestead; sugar, for all practical purposes, cannot.

Before we discuss making jams and jellies with honey we want to be honest and tell you that they are more difficult to make than the conventional jams and jellies made with sugar. Hopefully, you'll look on them as a challenge, a real test of your creativity and skill. When your jams and jellies are sparkling in their jars, you can take pride in the fact that you have made jellied products that are not only unusual, but natural and nutritious as well.

Many authorities are in disagreement about exactly how one should use honey in making jellied fruit products. Although some recipes we found indicate long cooking times for the jellies or jams, there seems to be a consensus among honey producers and users that high heat over long periods of time will destroy the biotic materials in honey and alter its flavor and color. This precludes using honey in cooked-down jelly recipes, and indicates that commercial pectins must be used in produce jellies or jams that are made with honey instead of sugar.

Using commercial pectins would not be a bad compromise at all if there were abundant recipes, or even suggestions, about how one could substitute honey for sugar in jelly recipes. There is much contradictory advice, however, and the best and safest bet is to do your own experimenting. All commercially made pectins include instructions in the package on how to use them, how to prepare the fruit, and when to add the pectin. Experiment with them, making small batches, and use honey instead of the sugar called for. You should be able to tell how much honey you'll need after only a few tests. Don't be discouraged by failures: we found that honey jams or jellies that have failed to gel make delightful syrups for homemade ice cream, cereals, and yogurt.

Raw Jams

In addition to cooked jellies and jams with pectin, you can make raw jams with fruit and honey. These jams are not cooked so they must be kept under refrigeration. Raw jams should be made in small batches and used up within a reasonable amount of time. They won't spoil because honey acts as a preservative, but their natural flavor will fade after several weeks. If you wish to make raw jams, we suggest that you freeze your fruit after harvest and thaw small quantities of it, a little at a time, to make raw jam as you need it.

MAKING JELLIES, JAMS, PRESERVES, CONSERVES, AND MARMALADES

There are certain categories into which jellied fruit products can be divided. There is a fine line separating jams, jellies, preserves, marmalades, and conserves, but fruit butters fall into a category of their own, so we will discuss them separately.

Jellies are cooked twice: first, the juice is extracted from the fruit (you can freeze the juice, if you wish, and make fresh batches of jelly at will), and then the juice is cooked with honey and pectin until it gels. Jams are purees made with fruit, honey, and pectin. Preserves, conserves, and marmalades are made with bits of fruit, cooked until translucent with honey and pectin. Preserves are generally made from a single kind of fruit (strawberry preserves contain only strawberries); conserves are made with fresh fruits and dried fruit or nuts, or both; and marmalades are made most often from one or many kinds of citrus fruit.

Cooking Your Fruit Mixture

Commercial pectin is commonly available in both powdered and liquid form. When using powdered pectin mix it with the unheated fruit juice for jelly, or with the unheated, crushed fruit for jams, conserves, or marmalades. If you are using liquid pectin, add it to the boiling juice and honey mixture when making jelly, and to the cooked fruit and honey mixture immediately after it is removed from the heat. When making jams, preserves, conserves, and marmalades all products should be brought to a full boil and cooked for the recommended length of time. A full boil is reached when bubbles form over the entire surface of the jams, marmalades, or conserves mixture, or when the jelly mixture reaches a full rolling boil that cannot be stirred down.

It is better to make a small quantity of jelly or jam at a time rather than a large one. Since honey foams when it boils, a large kettle is required. An eight or ten-quart kettle with a flat, broad bottom will be fine. Do not use an aluminum pot, as acids in the fruit may react with the aluminum.

We suggest you use a light-colored, mild-flavored honey for jams and jellies unless your family likes the taste of darker honeys. Dark honeys will impart a strong flavor to the product in which they are used, and although the taste will not be unpleasant, it probably will be too strong for most people. Clover, alfalfa, or another very light honey is best for making jellied fruit products. The honey should, if possible, be fresh because the flavor of honey changes somewhat in storage.

Filling and Sealing Containers

Jellied fruit products may be sealed with paraffin in jelly glasses (straightsided containers without inner lips) or they may be sealed in glass canning jars. Although jellied foods must be cooked in an open pot, they do not need to be processed in either boiling water or pressure steam like vegetables, meats, and other fruits that are stored in canning jars. You may seal jellied products with paraffin only if they are firm. Fruit butters, preserves, conserves, and marmalades are usually too soft to be sealed with paraffin. In warm, humid climates, where there is a chance that paraffin might melt, seal all jellied products in canning jars.

Get glasses or jars ready before you start to make the jellied product. If you are using canning jars, make sure all jars and closures are perfect. Discard cracked or chipped jars. Wash the containers in warm, soapy water and rinse with hot water. Keep them hot by placing them in a slow oven or in hot water, so that they will not crack or break when hot liquids are poured in them. Wash and rinse all lids and bands. Metal lids with sealing compound may need boiling or holding in water for a few minutes—follow the manufacturer's directions. If you are using porcelain-lined zinc caps, have clean, new rings of the right size for jars. Wash the rings in warm, soapy water and rinse them well.

Work quickly when packing and sealing jars. To keep fruit pieces from floating to the top, gently shake jars of jam occasionally as they cool. If you are using jars with two-piece lids, fill hot jars to one-eighth inch of the top with hot fruit or jelly mixture. Wipe the rim clean, place a hot metal lid on the jar with the sealing compound or rubber next to the glass, screw the metal band down firmly, to seal, and stand the jar upright to cool. If you are using jars with porcelain-lined zinc caps, place a wet rubber ring on the jar, fill the jar to one-eighth inch of top, screw the cap down tightly to seal, and stand upright to cool.

If you are sealing your jars with paraffin, remember that you can only seal firm products with it. Use only enough paraffin to make a layer one-eighth inch thick. A single thin layer can expand or contract readily and makes a better seal than one thick layer or two thin ones. Prick air bubbles in paraffin; they cause holes as the paraffin hardens and may prevent a good seal. It is best to use a double boiler to heat your paraffin. Heating it over direct heat is dangerous because wax is flammable.

Allow paraffined products to stand overnight before moving them about so that the wax can harden completely. Cover paraffined glasses with metal or paper lids. If you make more than one batch in a day, label each jar with the type of jellied product, the day it was made, and its batch number.

For best quality, do not keep jellied fruit products for more than six months. Like many canned goods, they lose flavor in storage. They should be stored in a cool, dry, dark place.

Extracting the Juice for Jellies

Before you make jelly, you must first extract juice from your fruit. To extract juice, prepare the fruit as directed in the recipe and then put it in a damp jelly bag or fruit press. The clearest jelly comes from juice that has dripped through a jelly bag without pressing. If you're not too concerned with producing a perfectly clear jelly, you can get more juice by twisting the jelly bag slightly and squeezing the juice out, or by using a fruit press. Pressed juice should be restrained through a damp jelly bag or a double thickness of cheesecloth without squeezing.

Jelly Tests

Some of our recipes call for a jelly test. There are two ways you can make a jelly test. To make a spoon or sheet test, dip a cold metal spoon in the boiling jelly mixture. Then raise it at least a foot above the kettle, out of the steam, and turn the spoon so that the syrup runs off to one side. If the syrup forms two drops that flow together and fall off the spoon as one sheet, the jelly should be done. Remember that honey jellies will be slightly softer than jellies made with sugar.

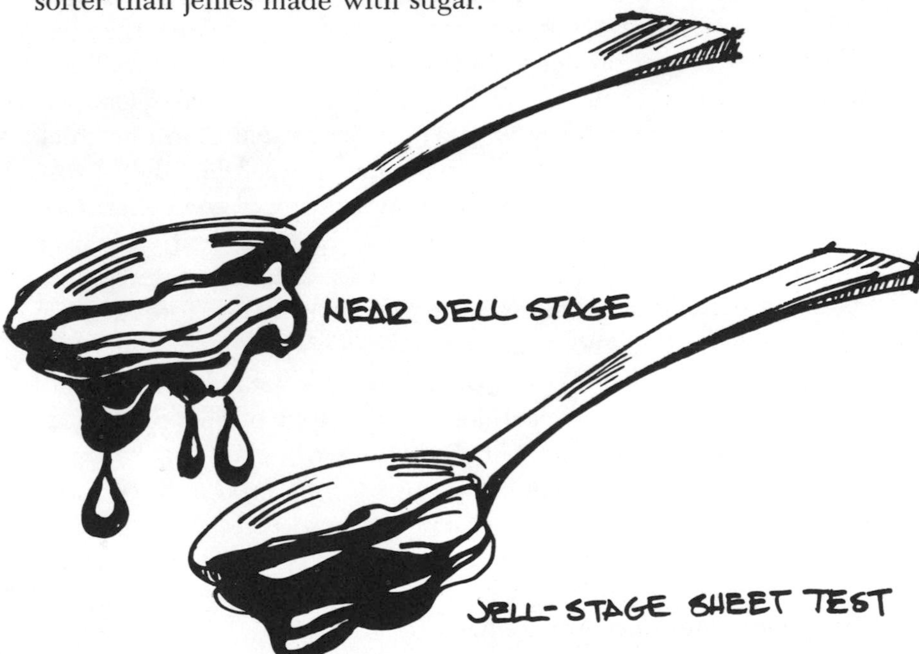

Continue to cook your jelly until it slides off a spoon in one sheet, without dripping or breaking.

You can also use the refrigerator test for your jelly. Pour a small amount of boiling jelly on a cold plate and put it in the freezer compartment of a refrigerator for a few minutes. If the mixture gels, it should be done. While you're testing the jelly remove the cooking jelly from the heat.

WHY DID MY JELLIED FRUIT PRODUCTS FAIL?

There are many factors involved in making jellied fruit products. It is often hard to pinpoint one single factor that is responsible for a poor jelly, especially if you have little experience making jellies with honey.

If you are using honey and your products are too runny, try using less honey next time. (Runny products can, as suggested before, be used as toppings for ice cream, yogurt, and cereal.) Honey jellies or jams tend to be softer than jams or jellies made with sugar. If, on the other hand, your jelly is too stiff, you may have used too much pectin. (Perhaps some underripe fruit in the batch contributed extra pectin to the product.) If the jelly is gummy, it may have been overcooked.

If jelly is improperly sealed, fermentation may occur or mold may develop. Jellied fruit products made with honey are generally a bit darker than those made with sugar; however, if your jellies or jams darken at the top of the jar, it might be because you stored them in too warm a place or because a faulty seal is allowing air to enter the jar. The color of jellied products can fade if they are stored in too warm a place or if kept in storage for too long a time. Red fruits, such as strawberries or red raspberries, are especially likely to fade.

If you've stirred your jams and preserves properly before jarring them, you should have no trouble with fruit floating to the surface. If you find fruit floating in your preserves, stir the preserve mixture gently for five minutes after removing it from the heat the next time you make preserves. Also, make sure that you've used fully ripe fruit, that it was cooked long enough, and that it was properly crushed or ground.

JELLY RECIPES

Honey Jelly

3 cups honey
1 cup water
½ bottle fruit pectin

Measure honey and water into a large kettle and mix. Bring to a boil over hottest heat and at once add pectin, stirring constantly. Then bring to a full rolling boil and immediately remove from heat. Skim; pour quickly into clean, hot, sterilized glasses and paraffin at once. Makes 6 glasses.

Lemon Honey Jelly

¾ cup lemon juice
2½ cups honey
½ cup liquid fruit pectin

Combine lemon juice and honey. Bring to a full rolling boil. Add pectin, stir vigorously and boil about 2 minutes. Pour into hot, sterilized glasses. Cover with paraffin to seal. This makes about 6 glasses of jelly.

Currant Jelly

Pick over currants. Do not remove stems. Wash. Drain. With a potato masher mash a few in bottom of preserving kettle. Continue adding and mashing currants until all are used. Add ½ cup water to 2 quarts fruit. Slowly bring to a boil and let simmer until currants appear white. Strain through a colander, then allow juice to run through a cloth or jelly bag. Measure juice. Add ¾ cup honey for every cup of juice. Boil until a good jelly test is obtained. Pour into hot, sterilized glasses. Cover with paraffin.

Apple Jelly

Wash apples. Remove stems and dark spots, quarter but do not pare or core apples. Add just enough water to half cover apples and cook until the fruit is soft. Drain, using a jelly bag. Measure juice. Add ¾ cup honey for every cup of juice. Boil until a good jelly test is obtained. Pour into hot, sterilized glasses. Cover with paraffin.

Variations of Apply Jelly

Mint Jelly—Just before removing apple jelly from the flame, add a few mint leaves which have been washed (about ¼ cup of mint leaves to 1 quart of juice) and a bit of green vegetable coloring. Stir, remove the leaves, pour jelly into hot, sterilized glasses. Cover with paraffin. This makes an attractive and delicious jelly to serve with lamb.

JAM, CONSERVE, PRESERVE, AND MARMALADE RECIPES

Making these products is easier than making jellies because only one cooking operation is involved. Everything is done in one pot. Be careful, however, to stir continually, because these products burn easily.

Peach Jam

4 pounds fully rippened fresh peaches
¼ cup fresh lemon juice
6 cups mild-flavored honey
1 package (3½ ounce) powdered fruit pectin

Wash, peel and remove pits from fresh peaches. Chop or coarsely grind peaches, blending with lemon juice. Measure prepared fruit, packing down in cup. You should have 4 full cups. Place fruit and lemon juice in large 6 to 8 quart saucepan. Add pectin, mix well. Place over high heat. Bring to boil, stirring constantly. (Oil rim of saucepan well and mixture will not boil over.) Oil measuring cup, measure honey. When fruit is boiling, stir and slowly pour in honey, blending well. Continue stirring and return to full rolling boil. When boil cannot be stirred out, boil exactly four minutes. Remove from heat. Alternately stir and skim for five minutes to cool slightly and keep fruit from floating. Pour into prepared glasses, allowing ⅛-inch for paraffin. Makes about 10 six ounce glasses.

Strawberry Jam

4½ cups prepared fruit (about 8 cups fully-ripened whole strawberries)
1 package (3½ ounce) powdered fruit pectin
7 cups mild-flavored honey

Gently wash strawberries in ice water. Drain. Hull, slice and crush 8 cups. Measure 4½ cups crushed fruit in very large saucepan. Add powdered pectin to fruit. Mix well. Place over high heat. Bring to a full boil, stirring constantly. At once, pour and stir in honey. Bring to a full rolling boil. Boil hard 2 minutes, stirring constantly. Remove from heat. Alternately stir and skim for five minutes to cool slightly and keep fruit from floating. Ladle quickly into prepared glasses. Cover jam at once with ⅛-inch of hot paraffin. Makes about 11 six ounce glasses.

Apricot Jam

3 cups dried apricots
¼ cup fresh lemon juice
7 cups mild-flavored honey
½ bottle liquid fruit pectin

Mix apricots, lemon juice, and honey. Simmer until apricot pieces are soft. Stir to prevent scorching. Remove from heat and stir in pectin. Stir and skim for 5 minutes. Pour into hot jars, and seal with lids. Makes 3 one-half pints.

Note: You can use dried peaches instead of dried apricots.

Raw Blackberry Jam

Mash ripe blackberries and add the amount of honey needed for your family's taste. Stir it in and keep the raw jam in a covered dish.

Blueberry Jam

In the blender put the following and blend into mush:

2 cup fresh or thawed blueberries
1 stalk rhubard
½ cup nuts

Pour the jam into a dish which can be covered tight. If too thin, stir in rice polishings to thicken and add honey to taste. Comb honey is better than the liquid type for raw jams. Keep refrigerated.

Part Cooked Jam

1 cup honey
1 box pectin powder
2 cups whole, raw cranberries

Slowly boil honey and pectin powder. Add ½ cup of cranberries. Chop up remaining cranberries in the blender and add to the jell after it has boiled for a few minutes. Chill.

Pineapple Jam

½ cup cold water
2 cups fresh pineapple
honey to taste
ground nuts
sunflower seeds

Put water and pineapple in the blender and whiz until smooth. Remove from blender and add honey, then ground nuts and sunflower seeds until you have a thick jam. Keep refrigerated in a covered dish.

Raw Strawberry Jam

Wash the desired amount of strawberries, then hull. Mash and add honey to taste. If the berries are very juicy, as they are some years, stir in either wheat germ flour or rice polishings, raw, and add enough to thicken the jam as you want it. Tiny new mint leaves may also be shredded up in the jam for an extra-special flavor. Keep refrigerated in a covered dish.

Tutti-Frutti Jam

Mash the following fresh fruits:

½ cup red raspberries
½ cup black raspberries
½ cup blueberries
1 ripe avacado
2 very ripe bananas

Add honey to taste, and thicken if necessary with rice polish. Add a pinch each of cinnamon and allspice. Keep refrigerated in a covered dish.

Sour Orange Preserves

The fruit of the native sour orange, so generally used for root stock over many portions of the citrus area, is used for making delightful preserves that are always popular.

For best flavor use the fruit when well matured and highly colored. Grate off all oil cells leaving the rich, yellow colored skin exposed. Cut into quarters and remove from pulp. Soak the peel in salt water (1 cup salt to 1 gallon water) overnight. Squeeze juice from pulp and save to add to preserves during the last cook. Drain peel from salt water. Cover well with clear water and boil for 10 minutes. Drain and cover with fresh water and cook until peel is tender. If no bitter flavor is desired, it may be necessary to change the water several times. However, if the fruit used is fully ripe the slightly bitter flavor is agreeable to most palates.

Drain peel and drop into a hot syrup made of three cups honey and two cups water for each 2 pounds of peel. Cook until peel is clear and syrup somewhat thickened. Remove from heat and let stand overnight. The next day, take from syrup, add ¾ cup honey and ½ cup sour orange juice and bring to boil. After boiling 10 minutes or until thickened, replace fruit. Boil another 10 minutes or until syrup is thick. Pack into hot jars immediately and process pints for 10 minutes at boiling. Grapefruit, tangelo, and shaddock peel may be preserved in the same manner as the sour orange.

Pear and Ginger Conserve

- ½ pound green ginger scraped and chopped
- 6 pounds honey
- 8 pounds pears weighed after paring and coring
- 1 pint water
- 4 oranges
- 3 lemons, juice and thinly shredded peel
- 2 cups pecans or black walnut meats

Cook the ginger, orange and lemon peel with a pint of water until tender, then add honey, orange and lemon juice; cook, put in the pears chopped coarsely and cook until pears are tender. Add nut meats. Cook 5 minutes longer. Pour in small hot jars and seal, boiling hot.

Pear and Cranberry Conserve

- 6 ripe pears, cored
- 2 cups frozen cranberries
- honey to taste
- pinch of cinnamon and cloves

Put pears and thawed cranberries in the blender. Whiz them to a fine pulp. Add honey, cinnamon and cloves. Whiz again. Keep refrigerated in covered dish. Red raspberries may replace the cranberries. Nuts or sunflower seeds may be added.

Raw Pear Conserve

In the blender (or through the food grinder) put these ingredients:

2 cups diced pears
½ cup organically-grown unsulphured raisins
½ cup honey
½ cup nut meats
¼ cup pineapple

Blend until you have a thick jam. Put in tightly covered glass jars and refrigerate.

Apricot Conserve

Pit and mash fresh apricots. Stir in the desired amount of honey and thicken with blanched, ground almonds.

Red Pepper Marmalade

12 medium-sized red peppers
2 cups chopped onions
2 cups white vinegar
3 cups honey
4 teaspoons salt
1 lemon, sliced
4 teaspoons whole allspice
½ teaspoon ground ginger

Remove stems and seeds from peppers. Cover peppers with boiling water; let stand 5 minutes; drain. Repeat, and drain well. Put through coarse blade of food chopper. Should measure about 4 cups. Tie spices in cloth bag. Combine with other ingredients. Boil 30 minutes, stirring occasionally. Let stand overnight. Next day, bring to boil in large saucepan and simmer 10 minutes. Ladle boiling hot into sterilized ½ pint pars. Seal. Makes about 6 half pints.

Honey Orange Marmalade

3 medium-sized oranges
1 cup water
1¾ cups honey
6 tablespoons lemon juice
¼ cup liquid pectin

Run oranges through food chopper, using the fine knife. Measure to make sure that you have at least 1¾ cups of ground orange. Add water.

Bring to a boil and simmer 15 minutes. Add honey and simmer 30 minutes. Add lemon juice, then liquid pectin. Bring to a full rolling boil and boil 30 seconds. Remove from fire. Skim and stir for 5 minutes. Pour quickly into sterilized glasses and paraffin.

Raw Marmalade

To make raw marmalade, use fresh fruits, wash and crush, or crush frozen and thawed fruits.

Any of the following fruits can be made up this way, as well as combinations of two or more fruits: Peaches, plums, strawberries, raspberries, blackberries, cherries, currants.

For each cupful of fruit, use about ⅓ cup honey. If you have honey which is crystallized, here is a good place to use it up.

Thicken the marmalade with wheat germ flour, using only as much as is needed. Keep refrigerated and covered.

Orange Honey

Scrub the skin of an organically-grown orange, quarter it, and blend until it is fine. Add it to 3 cups of raw strained honey. Keep it in a covered can in the refrigerator between meals. This is good on waffles, pancakes and corn bread.

USING LOW-METHOXYL PECTIN

Low-methoxyl pectin is far easier to use than other commerical pectins because it does not require any sugar or other sweetener to achieve a gel. Regular jams and jellies gel through an interaction of pectin, acid, and sugar. Those made with added commercial pectin are not different in this respect from cooked-down jellies which gel with the aid of the natural pectin of the fruits. Low-methoxyl pectin differs from both the commercial and natural pectin in fruits because it requires calcium salts, not sugar, to form its gel. Euell Gibbons originally discovered low-methoxyl pectin through his brother, a diabetic, who was experimenting with it for use in his own diet, and reported on his own experiments in ORGANIC GARDENING AND FARMING. At that time, Gibbons' big problem was obtaining a small quantity of the low-methoxyl pectin, since it was commercially available in large quantities only. That problem has been solved for him—

and for all of us—since Walnut Acres began handling low-methoxyl pectin and the calcium salts used with it.

Using the low-methoxyl pectin and calcium for jelly-making is very simple. Here is how it is made:

Prepare your fruit or juice. Generally, jam fruit is simmered until soft, then put through a sieve to reduce it to a pulp and to remove skins and seeds. Fruit for jelly is usually crushed, simmered, then dripped through a jelly bag or several thicknesses of cheesecloth to obtain a clear juice. You can also use canned, bottled, or frozen juices. Measure your juice or fruit. Put it in a saucepan over medium heat and bring just to a boil.

Next measure ⅛ cup honey and ½ teaspoon of low-methoxyl pectin to each cup of fruit or juice, and mix them well in a mixing bowl. Pour this mixture into the freshly boiling juice or fruit, all at once, and stir, stir, stir until the honey and pectin are completely dissolved. Now add 1 teaspoon of the calcium solution (⅛ teaspoon calcium in ¼ cup water) for each cup of juice or fruit, and quickly stir until well mixed.

The jelly or jam is done and can be immediately poured into sterilized jars and sealed.

However, one can test this jelly before sealing to see if it has exactly the right texture by putting about a tablespoonful in a metal cup and chilling it in the refrigerator. Such a small quantity chills very quickly; and as soon as it is cold, it will be the texture of your finished jelly. If it is too stiff, add more juice to the pot; if it is too runny, add another teaspoon or so of the calcium solution. It is that simple. (Remember to remove your jelly or jam from the heat while you're making this test.)

This jelly and jam cannot be merely paraffined and set on the shelf. It must be sealed in sterilized jars with sterilized lids, for even the microorganisms that cause spoilage find these products good. The half-pint, straight-sided mason jars that seal with two-piece dome lids are ideal. Sterilize jars by boiling them for 10 minutes, then leaving them in hot water until they are used. Heat the jelly or jam just to boiling point and pour it, boiling hot, into the sterilized jars, immediately sealing with the sterilized lids. When opened for use, jar will keep perfectly well in the refrigerator for several weeks.

The beauty of using low-methoxyl pectin is, of course, that since sugar isn't necessary for gelling, you're free to add as little or as much sweetening as you want. You also don't have to worry about adding too much liquid (honey or fruit juice) because you can always adjust the gel of the mixture by adding more juice if it's too stiff or more calcium if it is too runny.

Low-methoxyl pectin and the calcium salts necessary for its gelling action can be ordered from Walnut Acres, Penns Creek, Pennsylvania 17862. Instructions on the use of the low-methoxyl pectin are included, but we think that the directions (above) are much clearer and easier to follow. Remember that sugar is *not* necessary for a gel, and that you can easily substitute honey as a sweetener.

FRUIT BUTTERS

You may find that fruit butters will become your favorite method of preserving fruits. A big advantage of fruit butters is that they can be made in very large quantities. They also aren't as delicate as jellies are: it isn't essential to get them off the stove at exactly the right moment for fear they will not gel. Rather, fruit butters will thicken naturally as they cook down, so you won't have to worry about testing your fruit for pectin or acid, or do a jelly test every other minute to find out if the mixture is thick or not. The test for thickness of a fruit butter is simple, fast, and easy: Put a dab of the butter on a plate, let it set for a few minutes. If the dab does not separate, that is, if you don't notice liquid at the edges of the drop, the butter is ready to be put up.

However, there are disadvantages to making fruit butters. They have to be cooked for a long time—far longer than jams or jellies—before they are ready to be put up. During the entire cooking time, they must be watched closely and stirred frequently so that they don't burn or scorch. Scorched fruit butter is next to useless, and there's few worse feelings than burning twenty quarts of it. You'll have to make fruit butter when you can spend the entire day in the kitchen so that you can stir it frequently.

Although fruit butters made with sugar have been known to keep for years without spoiling (bacteria are too smart to eat the sugar which we, in our ignorance, consume), it is still best to *can* your butters. You can do this by processing both pint and quart jars in a boiling-water bath for ten minutes. (See section on canning fruits for directions for the boiling-water bath process).

Don't allow these disadvantages to dissuade you from making at least some fruit butters—they have a long and honorable history. In German, the word *Latwerg* orginally meant prune or pear butter which the people of the Rhenish Palatinate made to keep their prunes and pears over the winter months. It often took up to forty-eight hours to prepare these fruit butters, but the time was spent in celebration as the preparation of the butter became the occasion of a folk festival. The descendants of these Germans, the Pennsylvania Dutch, brought *Latwerg* with them when they came to America. In this country, the term came to be associated with apple butter because the Pennsylvania Dutch made huge quantities of apple butter in the fall, stirring apples and cider in huge kettles over an open fire. As you might imagine, this, too, was a festive event, and often used as an "excuse" for courting. Like many other once-regional specialties, apple butter has been made available to most parts of the country by modern transportation and improved marketing facilities; but store-bought spread just can't match the taste of real, honest-to-goodness homemade apple butter.

Equipment

A large kettle or pot with a heavy bottom is best for making fruit butters. Pennsylvania Dutch families used cast iron pots slung on tripods over outdoor fires. If you're making large quantities of fruit butters, you may wish to invest in a large stainless steel stockpot. (These pots also come in handy when you're making soups, sauerkrauts, and other foods in large quantities.) Stockpots come in various sizes, from small to huge, and you can choose according to your taste and budget. Many restaurant supply stores will give you a good price on a second-hand stockpot. Aluminum stockpots are considerably cheaper than those made from stainless steel, but don't be tempted by the lower price unless you plan to use it only for boiling-water bath processing. The acids in fruits may react with the aluminum and cause both a color and flavor change in your butters. Stainless steel is best because the metals in stainless steel are stable and will not react with acids in fruits and other foods. You may find that the bottom of stainless steel stockpots are thin and do not conduct heat evenly. If this is the case with your pot, buy an asbestos pad for the burner and place it under your pot when you are making fruit butters. This pad will spread the heat evenly over the bottom of the pot and will help to prevent hot spots from developing.

Some recipes recommend "baking" fruit butters in the oven. Again, stainless steel is good for this, and you may be able to buy a stainless steel pan from the same place you got your stockpot. Glass casserole dishes are good, too, if they are deep and wide (it is better to have a shallow layer of butter in a pan rather than a thick one—it cooks up faster). Enamel roasting pans are also good.

Adding Honey to Your Butters

You will notice that in most of the recipes that follow, we recommend adding honey to taste when the butter has finished cooking down, rather than adding honey while the fruit is still cooking. If fruits are ripe or slightly overripe, the natural sugars in them are at their peak, and no extra sweetener may be necessary. You may spoil your butter by making it too sweet if you add honey before tasting the finished product.

The ingredients in most of the recipes below are given in parts rather than in actual measurements, like cups or pounds. We have done this so that you can easily adjust the quantities and use just the amount of fruit you have on hand.

Apple Butter

Some people feel that the best apple butter is made from unpeeled, uncored apples. Other people feel that there's little difference in taste between butter made from apples that were peeled and cored before they were cooked down, and butter made from apples that were cooked whole and then put through a food mill or fruit press to remove the skins and seeds. The choice is up to you, but we offer a little advice: If you're not concerned about the slight difference in taste in the finished butter and want to save yourself a little time, peel and core your apples first, because it's more work to remove the skins and seeds later on by using a food mill or fruit press.

The recipe below is given for those who are using unpeeled and uncored apples. If you are peeling and coring yours, leave out the step that tells you to run your fruit through the fruit press.

3 parts cider
2 parts apples, unpeeled and uncored, sliced thin

Put the cider in a big pot and bring it to a boil. Add the apples slowly, being careful not to splatter yourself when you do. Allow the apples and cider to come to a boil, then simmer, stirring frequently to prevent sticking (and do remember to stir).

When the apple butter has begun to thicken considerably, the apple slices will start to fall apart as you stir the butter. At some point after the butter has thickened, remove it from the heat and put everything through a fruit press, discarding the peels, seeds, and stems. (If you have peeled and cored your apples, omit this step.) Put the remaining soupy mixture back in the pot, put the pot back on the heat, and simmer until the apple butter is a thick, dark brown mass. (One old cookbook said, ". . . til the liquid becomes concrete—in other words, till the amalgamated cider and apple become as thick as hasty pudding.")

Test to see if the apple butter has thickened. If it has, you may want to sweeten it. Use honey, of course, and sweeten to taste. You may also want to add spices to your apple butter. If so, add cinnamon, allspice, and ground cloves to taste. Bring the mixture to a boil, and jar the apple butter in hot, sterilized jars, leaving a ¼-inch headspace. Screw the lids on tightly.

Process pints and quarts for 10 minutes in a boiling-water bath. Remove from heat and seal jars if necessary. Cool them on a wire rack and store cooled jars in a cool, dark place.

Pear Butter

1 part water
2 parts apple cider
3 parts pears

Follow above recipe.

Grape Butter

1 gallon grapes
4 tablespoons water

Put grapes in kettle with water. Heat and mash the grapes. Continue cooking as the mixture thickens, stirring frequently. When thicker, put through a fruit press and remove skins and seeds. Return to heat and cook until thick.

If you want to sweeten your grape butter, add honey to taste. Add honey just before canning.

When thick, pack in hot, sterilized jars, leaving a ¼-inch headspace. Screw the lids on tightly. Process pints and quarts for 10 minutes in a boiling-water bath. Remove jars from canner and complete seals if necessary. Cool on a wire rack.

Peach Butter

3 parts sliced peaches
1 part water

Peel peaches and remove pits. (For freestone peaches, scald for 1 minute to remove the skins.) Place peaches in large pot and cook until peaches are soft. *Shake* pot frequently to prevent sticking (it is better if you divide your peach butter up among several pots for this part of the operation). When the fruit is tender, put it through a mill or a fruit press. Turn the purée into a shallow roasting pan or pans and cook uncovered for 1 hour at 325° F., and continue cooking, stirring every 15 to 20 minutes, until the butter is thick, fine-textured, and a rich reddish-amber color.

When it has reached this state, take it out of the oven. Ladle it into hot, sterilized jars. Screw the lids on tightly. Process pints and quarts 10 minutes in a boiling-water bath. Remove jars from canner and complete seals if necessary. Cool jars on wire rack and store in a cool, dry, dark place.

Apricot Butter

Cut fruit in halves, remove pits, and place in large kettle with enough water to prevent scorching. Cook until tender. When tender, put through a fruit press or press through a sieve. Cook until thick, stirring frequently to prevent scorching. When the butter is thick, sweeten with honey if you desire.

Pour into hot, sterilized jars. Screw the lids on tightly. Process pints and quarts for 10 minutes in boiling water bath. Remove jars from canner and complete seals if necessary. Cool on wire rack and store in a cool, dry, dark place.

Plum Butter

Wash plums and remove all blemishes. Put in kettle and add enough water so that the plums are just covered. Cook until tender. Put through food mill or fruit press to remove pits and skins. Measure pulp and add ½ cup honey for each cup of plum pulp, if desired. Return to heat and cook until thick. Pack in hot sterilized jars leaving a ¼-inch headspace. Screw lids on tightly.

Process pints and quarts 10 minutes in boiling water bath. Remove jars from canner and complete seals, if necessary. Cool upright on wire rack and store in a cool, dry, dark place.

Prune Butter

1 pound prunes
½ cup white or cider vinegar
honey to taste

Rinse prunes. Cover with water, bring to a boil, and then reduce heat and simmer until tender. Cool slightly. Remove pits. Put the prunes through a fruit press. Add the vinegar. You can add spices to the butter now, if you wish. Recommended spices include: ¼ teaspoon grated nutmeg, ½ teaspoon powdered allspice, ¼ teaspoon powdered cloves. Cook until thick and then sweeten with honey.

Pour into sterilized, hot jars leaving a ¼-inch headspace. Screw lids on tightly and process in boiling water bath for 10 minutes. Remove jars from canner, complete seals, if necessary, and cool upright on wire racks. Store in a cool, dry, dark place.

JUICING YOUR HARVEST

FRUIT AND VEGETABLE JUICES

Fruits and vegetables too bruised or overmature for canning or freezing need not be thrown away. Even though the skins outside of these foods may be in pretty poor shape, the juice locked inside the plant cells can be salvaged, and then canned or frozen for drinking, or using in soups, desserts, and casseroles later in the year.

It's true, of course, that whole fruits and vegetables have a proportionately greater food value than their juices; this is because the fiber, pulp, and skin are intact. However, by making juices from less than perfect produce, you're capturing much of the vitamins, minerals, and enzymes that would otherwise be lost if the whole foods were left to spoil. And for the gardener with storage problems, juicing foods might provide a solution. The juice from a pound of spinach will fill a small juice glass half full, so you can store a garden full of excess produce in a small place if you preserve it in the form of juice.

Using Juicers

Nowadays, a good selection of juicers is available. There are two basic types of juicers. The most common is comprised of a perforated metal basket that sits atop a high-speed rotary plate. The plate has hundreds of small raised cutting edges. When a carrot, for example, is forced into the hole on top of the machine, the teeth on the plate rip open the tissues and destroy the cell walls; juice and pulp are then free to go their separate ways. The high-speed whirling plate hurls the pulp and juice mixture at the walls of the basket. The pulp is retained by the basket, and the juice sprays off from the centrifugal force and runs down the case into a spout, where you collect it.

This is the usual type found in the home. It has to be cleaned out after each use, as the basket fills up with fine, slightly moist pulp. The second type has a screw-type cutter to rupture the cells, the juice falls by gravity, and the pulp is continuously ejected, so that you don't have to clean it during a day's juicing. This type is more suited to restaurants and health bars. It also costs more.

The following recipes come from Brownie's—the famous health food restaurant in New York City that started in 1936 as a vegetable juice bar. Brownie's still knows the value of these drinks and has prefected some truly delicious ones. In all these recipes, prepare the other juices first, pour them in an eight-ounce glass and fill with carrot juice.

Honi-Lulu

2-in. wedge pineapple
juice of ½ orange
carrots
Yield: 1 serving

Note: Citrus is better juiced in a citrus or hand juice squeezer.

Orange Blossom

½ McIntosh apple
juice of ½ orange
carrots
Yield: 1 serving

Sunshine

2-in. wedge pineapple
juice of ½ orange
dash of papaya syrup concentrate
squeeze of lime juice
carrots
Yield: 1 serving

Vegetable Garden

1 beet
3 leaves escarole
2 leaves chickory
1 scallion
1 radish
carrots
Yield: 1 serving

Juicing by Hand

But people have been making tomato and fruit juice for centuries without the aid of modern juicers. You need a little more patience if you are juicing by hand, but the operation is quite simple.

Use ripe apricots, apples, plums, any kind of berry, grapes, cherries, nectarines, peaches, pears, or tomatoes. The fruits may be smaller and less perfect than the ones you save for canning and freezing, but they must be very ripe. You only need to pit plums, peaches, and nectarines; the other fruits may be left whole, with skins remaining. Put the fruit in a large pot made of stainless steel, glass or enamelware. Add very little water, only enough to cover the bottom of the pot. Turn the heat up high, and as the fruit starts to simmer, cut through it with a sharp knife to release some of the juices. Cover the pot and allow the fruit to cook until tender. Stir occasionally to prevent sticking and add more water only if the fruit begins to stick.

When the fruit is tender, press it through two layers of cheesecloth or through a food mill or cone-shaped colander. Keep the pulp separate from the liquid you collect. The juice should not need sweetening because it contains natural sugars. If you do wish to sweeten it, add honey to taste. Usually one-half cup for each gallon of juice is sufficient.

Storing the Juice

If you are freezing the juice, pack it immediately after collecting in plastic containers or glass canning jars. Leave a little room for expansion if packing in glass. Freeze quickly.

For canning, bring juice to a rolling boil. Set the sterilized jars in a pan of warm water to prevent breaking and pour the boiling juice into the jars, leaving a one-eighth-inch headspace. Wipe the edge of the jars and adjust the rubbers and lids and seal. Invert the jars to sterilize the lids and let the jars cool in a place that is free from draft. Do not move the jars until they have cooled completely. If you're using jars with screw tops, remove the rings after twenty-four hours.

The pulp that you have collected may also be frozen or canned just as the juice, and can be used for fruit desserts and sauces.

APPROXIMATE NUTRITIONAL CONTENTS OF IMPORTANT JUICES

	BEET (Red)	CARROT	CELERY	CUCUMBER	Romaine LETTUCE	PARSLEY	RHUBARB	SPINACH	TOMATO	WATERCRESS	APPLE	COCONUT	GRAPE	GRAPEFRUIT	LEMON	ORANGE	PINEAPPLE	POMEGRANATE
PROTEIN %	1.6	1.1	1.1	0.8	1.2	3.5	0.6	2.1	0.9	1.7	0.1	1.4	1.3	0.4	0.9	0.6	0.4	1.5
FAT %	0.1	0.4	0.1	0.2	0.3	1.0	0.7	0.3	0.4	0.3	0.2	12.5	1.6	0.1	0.6	0.1	0.3	1.6
CARBO-HYDRATE %	9.7	9.3	3.3	3.1	3.0	9.0	3.8	3.2	4.0	3.3	12.5	7.0	19.2	9.8	8.7	13.0	9.7	19.5
CALORIES PER PINT	220	217	89	84	94	283	115	115	112	109	250	700	462	200	210	265	207	472
CALCIUM %	0.140	0.225	0.390	0.050	0.345	0.350	0.220	0.390	0.055	0.785	0.035	0.120	0.055	0.105	0.110	0.120	0.040	0.030
MAGNESIUM %	0.130	0.100	0.140	0.045	0.065	0.160	0.085	0.250	0.065	0.170	0.040	0.100	0.045	0.045	0.045	0.055	0.050	0.020
POTASSIUM %	1.770	1.540	1.460	0.700	1.660	1.50	1.625	2.685	1.335	1.435	0.640	1.500	0.530	0.805	0.615	0.905	1.350	1.600
SODIUM %	0.485	0.385	0.645	0.050	0.100	0.200	0.125	0.445	0.060	0.495	0.055	0.180	0.025	0.020	0.030	0.060	0.080	0.250
PHOS-PHORUS %	0.210	0.205	0.230	0.105	0.140	0.130	0.090	0.230	0.145	0.230	0.060	0.370	0.050	0.100	0.055	0.090	0.055	0.050
CHLORINE %	0.290	0.195	0.665	0.150	0.395	0.090	0.180	0.330	0.145	0.305	0.025	0.600	0.010	0.025	0.030	0.025	0.255	0.068
SULPHUR %	0.090	0.110	0.140	0.155	0.130	0.120	0.065	0.180	0.070	0.835	0.030	0.140	0.045	0.050	0.045	0.050	0.045	0.040
IRON %	0.004	0.003	0.003	0.002	0.007	0.016	0.003	0.013	0.002	0.015	0.002		0.0015	0.0014	0.003	0.002	0.002	0.004
SILICON %	0.009	0.007	0.008	0.013	0.018		0.006	0.020	0.009		0.006		0.002			0.0007		
MANGANESE %	0.008	0.0005	0.0014	0.0013	0.0064	0.008	0.0013	0.0042	0.0012	0.0036	0.0003	0.0017	0.0001	0.0001	0.0002	0.0003	0.006	
COPPER %	0.001	0.007	0.001	0.016	0.0003	0.015	0.0005	0.001	0.0005	0.005	0.0008	0.0009	0.0005	0.0003		0.0008	0.0004	0.0005
IODINE Parts per billion	230	180		500	650			400	350	180							200	120

U. S. Dept. of Agriculture Chart

This chart, prepared by the United States Department of Agriculture, measures percentages of various nutritive elements in vegetable juices. The bulk of juice is, of course, water. Cabbages eaten whole are 97 percent water, for instance, so by juicing them you're retaining more than 97 percent of their nutritive elements.

Easy Fruit Juice

Helen and Scott Nearing started canning fruit juices on their Vermont homestead many years ago. Their method for "putting up" juices sounds so simple, and yet so successful, that we think it's worth more than a mere mention here. So, we quote from their book, LIVING THE GOOD LIFE (Schocken Books, 1954):

> The glass jars were sterilized on the stove. A kettle or two of boiling water was at hand. We poured an inch of water into a jar on which the rubber had already been put, stirred in a cup of sugar until it had dissolved (we used brown or maple sugar, or hot maple syrup), poured in a cup and a half of fruit, filled the jar to brimming with boiling water, screwed on the cap and that was all. No boiling and no processing. The raspberries, for example, retained their rich, red color. When the jars were opened their flavor and fragrance were like the raw fruit in season. The grape juice made thus was as delicious and tasty as that produced by the time-honored, laborious method of cooking, hanging in a jelly bag, draining, and boiling the juice before bottling. Our only losses in keeping these juices came from imperfect jars, caps or rubber. We found that two people could put up fifteen quart jars in twenty minutes.

MAKING APPLE CIDER

What Apple Varieties to Use

Apples for cider need not process the flawless perfection we seek in table fruit. Here you can use the blemished, the bug-marred, the runts and the "drops," and the otherwise unchoice, mixing all apple varieties together or using only what you have. USDA extension horticulturists tell us that highly flavored cider begins with a band of suitable varieties. Apples should be firm-ripe, but not overripe. Peak ripeness is indicated by characteristic fragrance and spontaneous dropping from trees. Green, undermature apples cause a flat flavor when juiced. Don't use apples in brown decay because their juice will ferment too rapidly for a prematurely "hard" cider.

For pressing, apples are usually separated into four variety groups—sweet sub-acid, mildly acid to slightly tart, aromatic, and astringent. The best-flavored cider comes from a blend of varieties from all four groups. Sweet sub-acid types, which usually make up the highest percentage used in a cider blend, include Baldwin, Rome Beauty, Delicious, Grimes Golden, and Courtland. Some mildly acid varieties, which make up the next largest proportion, are Jonathan, Winesap, Stayman, Northern Spy, York Imperial, Wealthy, Rhode Island Greening, and Melrose. Varieties that add aroma to the juice include Golden Delicious, McIntosh, and Franklin. The small quantities, consist of such crabapple varieties as Florence, Hibernal, highly astringent group, used only in Transcendent, Martha, and Dolgo.

Juicing Apples by Hand

When all the fruit has ripened on your apple trees, you are ready for cider-making. If you have no fruit press and only a little fruit, ordinary household gadgets will suffice. Apples can be cut and run through a food chopper, or blender, or crushed with a rolling pin on a chopping board. Catch all juice runoff in glass or enameled ware, *but not in an aluminum or any other unglazed metal* container, because contact with metal gives apple juice a bitter taste and a dark gray color.

Put your crushed pulp into a clean muslin sack—an old pillowcase will do—and squeeze out all the apple juice possible. Now pour this juice into clean glass jugs or bottles, cover with a cotton wool plug, and let stand at room temperature. The bottles should be filled to just below the brim. Be sure the plug is in not tightly, but snugly. This will allow proper fermentation without adding the airborne "vinegar-bug."

After three or four days sediment will begin settling on the bottom as fermentation beads rise to the top. If you wish only a mild, sweet cider this is the time to "rack off" the clear liquid from the sediment and store in cold place for immediate use. "Racking off" is done by inserting one end of a rubber pipe (about three feet in length) into the liquid, and siphoning at the other end with your mouth, as you would with a soda straw. As soon as you feel liquid in your mouth, pinch off this end with your fingers and insert into empty container which should stand well below the filled one. Naturally all this equipment should be scrupulously clean. Rack off only the clear liquid, do not disturb the sediment at the bottom.

If you prefer a cider with more zip and on the dry side, allow "must" to keep standing in warm room. In about ten days it will begin frothing and may foam over the top. Replace with new cotton plug, clean off sides and let frothing continue till fermentation subsides. If you have an airlock—a curlicue glass "cork" sold by homebrew suppliers—use it in place of the cotton plug. If not substitute three thicknesses of clean muslin tightly stretched over bottle opening and secured well around neck. Be sure you use strong, sound glass, like cider jugs, or the bottle may burst with increasing fermentation.

Since this process turns all available sugars in the "must" to alcohol, it stands to reason this brew will no longer be a sweet drink—it will become "dry" cider. And the longer it stands, the "harder" is becomes. Alcohol—which does not freeze—may be extracted by allowing this drink to freeze solidly—remove corks first—and then running a hot poker through the frozen contents till you reach the free alcohol and pour if off.

If you prefer the mild drink with the definite essence of apple unaltered, fermentation can be arrested by pasteurization at 165° F. Pour fresh into sterilized mason jars and process for thirty minutes in boiling water bath, then completely seal. (For more information about processing by the boiling water bath see the section on canning fruits and vegetables.) Strain before serving. To preserve by freezing, pour into glass receptacles and leave uncovered till solidly frozen. Baked enamel pots and pans may also be used for freezing.

Using a Fruit Press

With a bushel or more of fruit to grind, it is best to use a fruit press. Be sure it is a *hard*-fruit press, with a cutting cylinder that minces the toughest apples with no strain at all. The portable presses are hand-operated, though there are power-driven models too.

Gather all your fruit together and pick over for decayed or wormy specimens, especially if picked from the ground. Hose it down and allow to dry in sun for about three days, spread out on racks or benches. A bushel of apples, thoroughly squeezed, will yield three gallons of juice, so gauge your container needs accordingly. All jugs, bottles, and kegs should be scrupulously clean. You will also need a large tub—not metal—to hold crushed pulp; a smaller one to catch juice runoff, and a clean muslin sack for juice pressing of pulp.

Set up your fruit press outdoors to facilitate easier handling and subsequent cleanup. If raining, or too cold, work in garage or shed. Apples should not be used when rain-wet. The operation of the press is so simple and self-evident that no instruction will be needed. With two people at the task—one cranking the grinder, the other feeding the cutting hopper—a bushel of apples can be ground up in fifteen minutes. No need to peel or cut up fruit when using a hard-fruit model; the cutting cylinder takes the toughest apple whole in stride. But *be careful,* keep hands and fingers away from rotating blades. As the pulp receptacle fills up, transfer the mash into your tub and repeat till all fruit is ground up. Be sure the smaller vessel is placed where it will catch all escaping juice. Now fill the muslin sack with only enough pomace to fit the juicer well, which is usually built under the mash well. Crank presser handle till no more juice runs and repeat process with rest of pulp. All squeezed juice may be temporarily contained in a wooden tub or enamelware.

Some cider experts contend the finest cider comes from pomace which has been exposed to the air for twenty-four hours and ground again. It is spread about and turned once or twice for fullest possible absorption of oxygen. If you wish to put this theory to a test, do up half of your pomace this way and compare results.

The rest of the work is a pleasure. Pour the apple juice through a straining cloth into your containers. Fill each almost brimful, and let stand at room temperature at least three days before drinking. As you siphon off the clear liquid from the sediment into new bottles, be sure to replace the loss of the settled portion—always keeping fermenting cider jugs full.

TURNING YOUR APPLE CIDER INTO VINEGAR

The hard part of making apple cider vinegar is extracting the juice, and, if you've followed the directions above, you've already done that. The rest is so simple. It is just a matter of allowing the extracted apple juice to ferment past the stage of sweet cider and dry cider into the vinegar stage.

Pour the strained apple juice into a crock, water-tight wooden container, dark-colored glass jars, or jugs. If you don't want your fermenting juice to run all over the floor, you'd better leave ample headspace—about 25 percent of the container—for the juice to expand during fermentation. Cover your container with something that will keep dust, insects, and animals out, but air in. A triple layer of cheesecloth, clean sheet material, or a tea towel will do nicely. Stretch this material over your crock, jars, or whatever, and tie it tightly with string. Store your brew in a cool, dark place, like a basement or garage. Now sit back and wait while the juice does all the rest of the work. Fermentation will take from four to six months. After about four months, remove the cover and taste the vinegar. If it is strong enough for your liking, strain it in a triple layer of cheesecloth, pour it into bottles, and seal with caps or corks. If the vinegar is too weak, let it work longer, testing it every week or so, until it is strong enough for you. If, when finished, your vinegar is too weak add some store-bought kind. If it's too strong, dilute it with a little water.

The "scum" or layer that forms on top of the vinegar during fermentation is called the "mother." This is what you want to strain off when your vinegar is strong enough so that it stops working. You can save this "mother" to use as a starter for your next batch of vinegar. Just pour your apple juice—or any other fruit juice or wine—into a crock, wooden container, or dark glass jars, add the "mother" and let it ferment. In a little while you'll have vinegar.

It is a risky business pickling food with homemade vinegar because, unlike store-bought vinegar which has a controlled acid content, the acidity of homemade vinegar varies. If you wish to pickle with your own vinegar, we suggest that you make small batches at a time so that you won't stand the chance of ruining all your relish or chow-chow.

Herb Vinegars

While we don't advise pickling with homemade vinegar, there is nothing wrong with adding a few herbs and spices to your home brew. Herb vinegar is delicious and very easy to make. Just add individual herbs like tarragon or garlic, or a combination of your favorities to your vinegar.

Louise and Cyrus Hyde, owners of Well-Sweep Herb Farm, in Port Murray, New Jersey, make herb vinegar for friends, visitors to their organic farm, and of course, for themselves. Although almost any herb can be added to vinegar, they have found a few to be the most popular. The favorities are tarragon vinegar, basil and tarragon vinegar, and basil and garlic vinegar. Other popular kinds are dill vinegar (made from seeds and leaves); rosemary vinegar, a purple-tinted vinegar made from opal basil and sweet basil leaves; orange-mint vinegar for fruit salads; and cucumber-flavored salad burnet vinegar.

You can make herb vinegar simply by adding fresh sprigs of herbs to bottles of vinegar, as shown here. You may also use crushed dried herbs instead of the fresh ones, but because they are in more concentrated form, add smaller amounts of them to each bottle.

Because the Hydes like to have herbs floating in their bottled vinegars, they add fresh sprigs of herbs—about a handful or three or four sprigs—to each quart of vinegar. The mixture is allowed to set at least two weeks before it is used. The herbs stay in the bottle and are poured onto a salad along with the vinegar.

If you prefer herb vinegar without the leaves floating in it, crumble about three tablespoons of dried leaves in a jar. If you're adding herb seeds, such as dill or anise, crush them well first. Then warm your vinegar and pour it over the herbs. The warm vinegar decomposes the leaves and extracts oil from the herbs more quickly than cool vinegar. Let the vinegar and herbs set for two to four weeks in a covered bottle or crock. Stainless steel and porcelain containers are also fine, but never use any other metal because the acid in vinegar will react with it and give your herb vinegar an undesirable appearance and taste. After a few weeks, test your vinegar to see if it is flavorful enough for you. When its flavor is right, strain it into sterilized jars and cap until you're ready to use it.

The Hydes like the distinct flavor of apple cider vinegar, but find that when used alone with herbs, its strong flavor masks the more subtle flavor of the herbs. For this reason they prefer to mix apple cider vinegar with the milder-tasting white, distilled vinegar before they add herbs to it.

DAIRY PRODUCTS

FREEZING MILK AND CREAM

If you're keeping even just one milk cow or goat, there are probably times when you've got more milk on your hands than you and your family can consume daily. There are some very simple things you can do with the excess. You can open a roadside stand or sell it from door-to-door, you can give it away to friends, or you can store it for times when your animal's milk production is at its lowest. Here, we'd like to spend some time discussing the third alternative: storing milk.

Just how do you preserve this highly perishable food? One of the simplest ways is to freeze it. Milk may be frozen whole or skimmed, pasteurized or raw. Even cream can be frozen. To freeze, pour your milk or cream into glass jars or plastic containers, leaving two inches headspace for expansion. Seal tightly and place in the coldest part of your freezer so that it freezes quickly. Whole milk will keep safely in the freezer for four to five months; cream should not be stored frozen for more than two or three months. Both milk and cream should be thawed for two hours at room temperature before using. The only problem you will encounter when freezing cream is that the oil (butterfat) tends to separate out during freezing. For this reason, thawed cream has limited uses. It can't be used successfully as is, poured into coffee or over cereal or fruit. It may not whip properly, but it can be successfully used for frozen desserts, like ice cream. If you want to use it for cooking—to make creamed soups, gravies, custards, and the like—or for baking, beat it a little first just so the oil is not floating on top.

MAKING BUTTER

A time-honored way of preserving cream is to churn it into butter. Although butter is also a perishable food, it will keep longer than milk or cream under refrigeration if all the buttermilk is worked out of it. Because butter is a concentrated form of cream—one gallon of cream will yield about three pounds of butter—it takes up a lot less storage space than cream.

SEPARATING THE CREAM

The first step in butter-making is to separate the cream from the milk. This is easy to do if you're using cow's milk. The butterfat in cream is lighter in weight than the other ingredients in whole milk and will rise to the top naturally by gravity in twenty-four to thirty-six hours. There are two simple methods of separating the cream: The shallow-pan method and the deep-setting method. In the shallow-pan method, the milk is drawn from the cow and immediately poured into shallow tubs or pans. These pans or tubs are placed in a cool spot, like a refrigerator, basement or spring house, for at least twenty-four hours. The cream that has risen to the top is then skimmed off with a flat dipper. Although this is certainly the easiest way to separate cream, some farmers object to this method because during the time it takes for the cream to rise, the surface of the milk is exposed to the air and frequently absorbs or develops objectionable odors and tastes.

A more satisfactory method of separating cream is by the deep-setting method. By this method, the milk is drawn from the cow and immediately poured into cans or buckets which are placed in cold water or, preferably, ice water for twelve hours. The quick cooling of the milk causes the cream to rise more rapidly and more completely. The cream can be skimmed in half the time required by the shallow-pan method, and its freshness and sweet flavor are retained.

If it's goat's milk you're working with, the job is a little more difficult. The fat globules in goat's milk are small and well-emulsified which means that the cream will take much longer to separate out than cow's cream. If you let goat's cream take its time to separate out, it may begin to develop a strong "goaty" flavor that most people find unappetizing. To separate the cream properly and quickly a cream separator, which separates the cream from the skim milk by centrifugal force, is needed. Warm milk (between 80 and 90° F.) is poured into the separator where it is whirled around. The cream and skim milk are released through separate spouts in minutes, while both are still warm and fresh. Some cow owners like to use a cream separator instead of letting their cow's cream separate by gravity, because the separator removes almost every drop of the butterfat from the milk very quickly. The skim milk is fat-free and can be used immediately for drinking or for making cottage cheese.

A cream separator is a delicate piece of equipment and should be cleaned and operated according to the manufacturer's directions. For best results, milk poured into the separator should be warm (not below 90° F.) and fresh. Most separators are left in the milk room or other unheated area. In winter, the cold temperatures can chill the instrument, and if the separator is cold enough, it will cool the first milk that is poured in to below 90° F. To prevent this from happening, warm the separator by running warm water through it before pouring in the milk.

The separator should be cleaned and sterilized immediately after each use. All parts should be rinsed in warm water and then scrubbed with a brush, warm water, and soda ash or a cleansing powder made especially for use in dairies. Soap should not be used because it is difficult to wash off completely and may leave a soapy film on the equipment. All the parts of the separator should be sterilized in a farm sterilizer or in boiling water for five minutes.

Cream separators advertised in DAIRY GOAT JOURNAL retail for about 75 dollars. Used ones are cheaper, of course, but since the popularity of keeping goats for milk has increased in the last few years, the demand for used separators is growing and the supply is diminishing.

CHILLING AND RIPENING THE CREAM

It is not very practical to churn a few cups of cream at a time, so many farmers who have just one or two milk-producing animals will collect cream over a few days, waiting until they have enough separated to make churning worthwhile. This is fine, but don't hold cream for more than four days before making butter. Butter made from old cream has an acidic, overripe taste, and it spoils quickly. If you are collecting cream over a few days' time, keep its temperature below 50° F. and don't add to it any cream that is not cooled to at least this temperature. The addition of warm cream raises the temperature of the older cream and hastens souring. Mix all the cream together—after it is all chilled—twelve to twenty hours before you churn it, and stir it occasionally with a stirring rod (a smooth rod with a four to five inch diameter disk on one end) or long-handled spoon so that it will have a uniform thickness.

The best way to cool cream rapidly is to cool it in a tub of ice water. Ice water will cool the cream more quickly than will refrigerator temperatures, providing the water is allowed to circulate on all sides as well as under the cream container. If ice water is not available, use plain water, but change it frequently. If you are fortunate enough to have a stream or spring nearby, put your container of cream in it. Its flowing cold water will do a fine job of cooling your cream. Pick a shaded spot—out of direct sunlight—set down your cream container and cover it with a clean towel or piece of cheesecloth to keep out dirt, insects, or other contaminants.

When the cream has reached a temperature of 50° F. it can be placed in a cool spot until churning. (Get yourself a floating dairy thermometer, available for under 2 dollars at farm supply stores, to measure temperatures accurately.) All cream should be kept at 52 to 60° F. in the summer and 58 to 66° F. in the winter while it is being churned. If the cream is too warm when it is churned, the butter develops too soon and is too soft and greasy. If the cream is too cold when churned, not all of the butterfat will separate out to form butter. This results in creamy buttermilk and less butter.

Cow's cream which is allowed to ripen before it is churned will produce more flavorful butter than that which is made from sweet cream. You can ripen raw cream by allowing it to set at room temperature (65 to 75° F.) until it is thick and slightly sour. To ripen milk quickly, add about one-half cup cultured buttermilk or yogurt. Fresh cream should not be added after ripening begins. Once ripened, the cream should be cooled quickly in a container of ice water until it reaches churning temperature. It should be kept at churning temperature for at least two hours before it is churned. Don't try to ripen goat's cream before churning. The goaty or cheesey flavor it will acquire will produce an unpalatable butter.

Churning

The old wooden upright churn with its long dasher is rarely used for butter-making anymore; most of these relics have found their way into antique shops. The wooden churn more commonly found in operation today is the barrel churn. Electrically powered and hand-operated glass churns are also popularly sold. If you want to make just a small amount of butter you don't even need a real churn. You can improvise with a blender, cake mixer, or even hand rotary beater. Early American settlers made butter by shaking cream in a deep wooden lidded bowl. You can use a glass jar. Pour cream into a jar until it is one-third full and start shaking. This is a rather tedious way of making butter and it calls for a strong arm, but it does work.

Whatever device you use to churn butter, make sure it is thoroughly clean before any cream is poured into it. If you have a wooden churn that is used only occasionally, it is advisable to fill it with water twenty-four hours before you plan to use it so that the wood will swell and be watertight. Scald the wooden churn with boiling water and then chill it down to churning temperature by filling it with ice water or placing it in a refrigerator, spring house, or cool basement before using. Glass churns, mixers, blenders, rotary beaters, and glass jars should be sterilized in boiling water and cooled before using.

Pour the cream into your churn, blender, or whatever equipment you are using, through a strainer to make sure that there are no lumps in the cream before you begin. Fill your churn only one-third full. Butter made from goat's cream is white. If you wish to color it, now is the time. Add a few drops of vegetable coloring to attain the desired shade of yellow. (Colonial housewives colored their butter with carrot juice.)

If you are using a hand-operated or electric churn, churn about ten times and then lift up the lid or remove the plug to permit gas to escape. Churn twenty times more and allow gas to escape again. Then resume churning, at about sixty revolutions per minute, until beads of butter about the size of corn kernels form. The churning process should take about thirty to forty minutes. Approximately twenty minutes will pass before you will hear the splash of the beads forming and feel the thickness of the butter.

When churning is finished, strain off the liquid. Don't throw it away. This is buttermilk. It won't be as thick as the commercial kinds because it is not cultured. It is lighter than regular milk and has a natural effervescence. It's the real, old-fashioned buttermilk that makes delicious pancakes, biscuits, and breads. What's left in the churn is mostly butter, with a little buttermilk mixed in. This remaining buttermilk must be removed to obtain the taste and texture of good butter. If it is not removed, the butter will have a shorter keeping quality and have a slightly acidic taste.

Wash the butter with clean water. Water temperature may vary according to the temperature of the butter that has formed, but it should be about 60° F. If your butter is too soft and warm, make your wash water cooler than 60° F; if it is too hard and cold, have your wash water a little warmer than 60° F.

The washing should be done right in the churn. Pour as much clean water as there is buttermilk into the churn after the buttermilk is poured off. Close the churn and churn it a few times to wash the butter. Pour off the cloudy water and repeat the washing process with fresh water. If this water is cloudy when poured off, wash again until the rinse water stays clear.

Now pour your butter into a large shallow bowl. Work out every drop of liquid by pressing and squeezing the butter against the sides and bottom of the bowl with a wooden paddle until

WHIPPING CREAM

1. To make small amounts of butter in a blender, fill the blender container no more than ⅓ full with cream and whip at lowest speed. 2. When yellow beads about the size of corn kernels form, pour off the buttermilk. 3. Rinse the remaining buttermilk out of the butter by pouring cool, clean water into the blender. Turn the blender on low for a few seconds to wash the butter. 4. Pour off the cloudy rinse water and repeat until the rinse water pours off clear. 5. Scoop the washed butter into a shallow dish and work out every drop of liquid by pressing the butter against the dish with a rubber spatula or wooden paddle as shown here.

no water can be poured off. Do not spread or thin the butter on the sides and bottom of the bowl; this makes the butter greasy.

If you're working with a home food blender, pour in the cream until if fills the container about one-third full. Set your blender at its slowest speed. Once the blades begin, remove the cap and watch the cream. It will first get foamy and then begin to thicken. Yellow beads will start to form in about four to five minutes. Once they get to be the size of a kernel of corn and the liquid seems a bit watery (like skim milk), your butter has formed. Turn off the blender and pour off the buttermilk. Pour in the wash water, turn on the blender for a second or two, pour off the water, and repeat the washing until the wash water stays clear. Pour the butter into a large shallow bowl and work all the liquid out in the manner described above.

If you're churning with a rotary beater or cake mixer, pour the cream into a deep bowl that has been sterilized with boiling water and chilled in the refrigerator. Use your mixer at its lowest speed. If you are beating with a rotary beater, whip at a constant speed which is comfortable for you. Do not stop beating until the butter has formed. Then pour off the buttermilk and add an equal amount of clean water. Beat for a second or two and pour off the wash water. Repeat until the wash water pours off clear. Then place the butter in a large shallow bowl and work out the remaining liquid.

If a glass jar with a tight-fitting lid is to be your churn, fill it one-third full with cream and shake about ten times. Then remove the lid to allow the gas to escape. Screw on lid and shake about twenty times more. Remove lid and let the gas escape again. Replace the lid and resume shaking without stopping until lumps of butter form and the liquid takes on a thin and slightly watery appearance. Pour off buttermilk and replace with fresh water. Shake jar about five times, pour off the wash water and wash again, until the wash water pours off clean. Place the lumps of butter into a large shallow bowl and work out the remaining liquid.

If your butter takes an unusually long time to develop or it just never comes at all, even after several hours of churning, obviously something is wrong, either with your equipment or your cream. The U.S. Department of Agriculture, in a farmers' bulletin (which is now out of print) describes some of the reasons why you may have problems forming butter:

1. The churning temperature is too low. Normally, it should be 52 to 60° F. in summer and 58 to 66° F. in the winter, but under exceptional conditions, it might be necessary to raise it to 65 to 70° F. This is especially true if your churn is very cold and you are churning in an unheated area on an exceptionally cold day.
2. The cream is too thin or too thick. It should be about 30 percent butterfat for best results.
3. The cream is too sweet. Very sweet cream will need to be churned longer than cream which has been ripened until it is thick and slightly sour. Ripening can be speeded up by adding about one-half cup of a starter, like *cultured* buttermilk or yogurt, to the cream.

4. The churn is too full. The churn should not be more than one-third full, no matter what type of equipment you are using for churning. The extra space allows the butterfat to move about freely.
5. Ropy fermentation of the cream preventing concussion. This may be prevented by sterilizing all the utensils and producing milk and cream under sanitary conditions. If additional measures are needed, pasteurize the cream, being careful to keep it from contamination after pasteurization. Then ripen the cream with a starter before churning.
6. Individuality of the animal. The only remedy is to obtain cream from a dairy animal recently fresh or cream that is known to churn easily. Before ripening it, mix it with cream that is difficult to churn.
7. The goat or cow being far advanced in the period of lactation. The effects may at least be partially overcome by adding, before ripening, some cream from another goat or cow that is not far advanced in the period of lactation.
8. Feeds that produce hard fat. Such feeds are cottonseed meal and timothy hay. Linseed meal, gluten feed, and succulent feeds such as silage and roots tend to overcome the condition and make churning the cream into butter easier.
9. Off flavor of butter. Influenced by cow's feed. Refrain from giving cows strong-flavored and strong-odored foods like turnips. If off-flavor persists, milk cows before, not after, feeding.

SALTING AND STORING THE BUTTER

Butter which is worked free of all its buttermilk and wash water may be eaten or stored as it is. This is unsalted, sweet butter. Salt, which enhances the flavor and lengthens butter's keeping quality, may be added at this time, if you wish. If you salt the butter, add ⅔ tablespoon of salt for each pound (2 cups) of butter. Work the salt into the butter by pressing and thinning the butter in the bowl with the wooden paddle, then adding a little salt and folding the butter over. Repeat this process until all the salt is worked in and the butter is firm and waxy. Don't spread the butter to thin it; this causes the butter to become oily and lose its firm texture.

When the butter is "worked," it is ready to be placed in appropriate containers and stored. You can roll your butter into a ball (or balls) and wrap it in aluminum foil or heavy-duty plastic wrap. You can press it into small bread pans and cover the pans with wrap. Or you can put the butter into glass jars with lids. It is not a good idea to store butter in plastic containers. These containers are porous and will allow air and strong odors to penetrate their walls. The taste of butter deteriorates the longer it is stored in the refrigerator. Keep it no longer than two weeks at refrigerator temperatures. If you wish to keep your butter longer than two weeks, freeze it at temperatures of 0° F. or colder. Do not keep it frozen for more than six months. Thaw butter for about three hours at refrigerator temperatures before using.

MAKING COTTAGE CHEESE

Cottage cheese got its name from the fact that it is a cheese that can easily be made in the home—or in the cottage, as the term once was. Traditionally, this soft, perishable cheese was made from the skim milk left after buttermaking. After the cream had been taken off the top, the remaining milk was poured into a good-sized crock and set in a warm place (usually on the back of the wood stove in the kitchen) for two days or so, until the milk had clabbered and much of the whey had separated from the curd. Then the curd was cut into cubes and heated gently over a low fire to firm it up a little. The warm curd and whey were poured into a cheesecloth sac and hung up over a tub until all the whey had drained off and only the soft curds were left. The cheese was then chilled for a few hours and then mixed with a little salt and fresh cream and stored in the spring house, cellar, or ice chest where it could be kept as long as five days without spoiling.

Cottage cheese made at home today is made in much the same way as it was one hundred or so years ago. Most of it is made with fresh raw skim milk, but fresh pasteurized or raw whole milk will do, too.

Raw milk contains the beneficial bacteria that are responsible for making milk clabber. Pasteurized milk has been heated to temperatures that kill this bacteria, and it will not clabber by itself, no matter how long it sits in a warm place. If you're using pasteurized milk, you'll need to add something to activate the curd and start the milk clabbering. You can add a milk product that contains the necessary beneficial bacteria, like cultured buttermilk or plain, unsweetened yogurt. (Natural buttermilk, the kind you pour out of the churn when you are making butter, won't help to make milk clabber. You'll have to use the thick, store-bought kind.) You may also use rennin, an extract made from the lining of unweaned calves' stomachs, which may be bought in the form of rennet tablets. These are the rennet tablets that you use to make rennet pudding or junket, not the kind you use to make hard cheese. They are available in most food stores.

In addition to the milk and activator, you'll need an earthenware crock, glass casserole dish, or stainless steel or enamel pot. Also have on hand a dairy thermometer, a long spoon, (preferably made from glass, wood, stainless steel, or enamel) a spatula or wide knife, a large pan or shallow pot that is larger than your crock, cheesecloth, and a colander. All your equipment should be scrupulously clean. Wash it with soap and water and rinse well with very hot water before bringing it in contact with the milk.

As we said before, cottage cheese is simple to make:

CLABBERING THE MILK

Pour a gallon of fresh raw or pasteurized, skimmed or whole milk into a crock or pot. If the milk is cold, bring it to room temperature over very low heat. If you are using pasteurized milk or if you want to speed up the clabbering process of raw milk, add one of the following:

4 tablespoons unflavored fresh yogurt

½ cup fresh cultured buttermilk

¼ table rennet (dissolved in ½ cup warm water)

Mix the milk and the activator together and cover the crock loosely with cheesecloth or a thin towel. Be careful not to smother the milk by covering it with a heavy towel or plate. All you want to do is to cover the milk to keep dust and insects out of it. If air isn't permitted to pass over the milk, the milk can acquire a musty odor and taste which will linger in the finished cheese. Let the covered crock sit in a warm (75 to 85° F.) place until clabbered. If you are using an activator, the milk should be clabbered in twelve to eighteen hours. Raw, inactivated milk will need about forty-eight hours to clabber.

CUTTING THE CURD

The milk has clabbered when much of the whey, a thin watery liquid, has risen to the top; and the curd, a white substance with a consistency similar to soft cream cheese, has settled to the bottom. Now it is necessary to cut the curd to allow more whey to separate out. If you are using skim milk, cut the curd into two-inch cubes. Do not make them much smaller than this, because if you do, too much whey may separate out, making the resulting cheese dry and leathery. If you are working with whole milk, the cubes may be cut smaller. To cube the curd, cut into two-inch strips with a spatula or wide knife. Then slice the curd again crosswise so that you have two-inch squares. If the curd is more than two inches deep, bring your spatula under the curd and cut across horizontally. It is important to cut the curd carefully so that just enough—but not too much—of the whey can separate out.

HEATING THE CURD

Pour a few inches of water into another large pan or pot, set the crock with the cubed curd in it and place it over low heat.

1. The first step in cottage cheese making is to clabber the milk. Pasteurized milk must be activated with buttermilk, yogurt, or rennet before it will clabber.

2. Once the milk has clabbered, cut the curd into 2-inch cubes to allow more whey to seperate out.

3. Heat the curd over boiling water until it reaches 115° F. If the curd is heated higher than this, the cheese will be tough and dry.

4. After heating, check the curds for firmness. They should hold their shape and be neither too soft nor too dry.

5. Pour the curds through a colander lined with cheesecloth to drain off all the whey, which otherwise would impart a bitter taste to the finished cheese.

6. You may rinse curds to wash off any remaining whey. To do this, place the colander with the curds in a bowl of water and stir the curds gently. Let the curds drain again before storing in the refrigerator or freezer.

Insert the dairy thermometer in the curd and heat until the thermometer reads 115° F. Hold it at this temperature for about half an hour, stirring occasionally so that the heat will be distributed evenly throughout the curd. Stir gently so as not to break the curd.

STRAINING THE CURD

After about thirty minutes, the curds will have settled to the bottom of the crock. Line a colander with cheesecloth and place it in a bowl to catch the whey. (Whey is an important source of minerals and B vitamins. Instead of pouring it down the drain, add it to soups, beverages, or casseroles, cook rice in it, or feed it to your pets or livestock.) Gently pour the curds and whey into the colander. Allow most of the whey to drain off and then take up the four corners of the cheesecloth and hang the curds in the cheesecloth over the bowl so as to catch the remaining whey that drips out. Let the curds hang this way until no more whey drips out. The curds may be rinsed at this time with clean cold water if you wish to minimize the acid flavor, although rinsing is not necessary. If you want to freeze your cottage cheese, do not rinse the curds.

STORING YOUR COTTAGE CHEESE

If you are freezing your cheese, take it from the cheesecloth without rinsing out the remaining whey, pack it in containers and freeze. If the cheese is to be eaten within a few days, it may be rinsed and should be chilled for a few hours. Then mix it with salt and fresh cream to taste if you wish. The cheese can be seasoned with herbs, like dill, chives, or parsley. If the cheese will be used in cheesecakes, blintzes, or pastries, do not add any seasoning or cream unless the recipe says to do so.

One gallon of milk should make about one and one-half pounds of cottage cheese.

MAKING YOGURT

Basically, yogurt is milk that has been fermented by special strains of beneficial bacteria. Yogurt can be made out of any kind of milk, be it raw or pasteurized, cow's or goat's milk, which has been skimmed or left whole. The starter bacteria that you introduce into the milk may be unflavored commercially prepared yogurt, your own homemade yogurt, or a pure culture from a natural food store.

Since there are several ways to encourage fermentation, you might like to experiment with a few methods first, and then choose the way that works best for you. Your choice will depend upon how much equipment you want to use and how much time you want to spend in yogurt-making. Whatever equipment you do use, make sure that it is thoroughly clean. Remember, you want to encourage the growth of beneficial and not harmful bacteria. An unclean bowl or spoon might give an off-taste to your finished yogurt.

MAKING YOGURT IN THE OVEN

Nutritionist Adelle Davis makes her yogurt in an earthenware bowl because it retains heat well. Miss Davis says to mix one quart of milk with one-fourth cup commercial yogurt. Fasten a cooking thermometer to the side of the bowl or float a dairy thermometer on the surface. Set the bowl in the oven and heat slowly to 120° F. Turn off oven, cover milk to hold in heat, and let cool gradually to 90° F. Maintain temperature between 90° to 105° F. by reheating oven if needed (about two to three hours) until milk becomes the consistency of junket. Check frequently during the last thirty minutes. Chill immediately after milk thickens.

USING A YOGURT-MAKER

Natural foods expert Beatrice Trum Hunter prefers to use a commercial yogurt-maker or culturizer. This yogurt-maker consists of a constant-temperature, electrically heated base and a set of plastic or glass containers with tight-fitting lids. Yogurt-makers make four individual pints or quarts at a time, depending upon the model. They are foolproof, cost from 10 to 15 dollars and are available at most natural food stores. Mrs. Hunter says to pour a quart of milk into a pot and bring to a near boil. Cool to lukewarm (105 to 115° F.). Mix the contents of a package of Bulgarian yogurt culture into the milk with a wooden spoon. Pour the mixture into pre-warmed cups of a yogurt-maker, cover, and leave undisturbed for about two hours. At the end of this time, remove the lid from one container and gently tilt the glass. The yogurt should be about the consistency of heavy cream. If it's still thin, let it incubate longer and check again. When the yogurt thickens, remove and refrigerate.

THE LONGEST, BUT SIMPLEST METHOD

There's also a way to make yogurt using no special equipment at all. It takes a little longer, but is just as easy and less expensive since you don't have to buy any thermometers or yogurt-makers.

Heat milk to almost boiling and let it cool. When the milk is lukewarm, add one-half cup yogurt for each quart of milk and stir well to make sure there are no lumps. Then warm a casserole dish by running hot water over it or heating it for a few seconds in a very low oven. Pour the mixture in the warm dish and cover. Wrap the casserole dish in a large towel and set in a quiet, warm spot in the kitchen. Leave undisturbed for at least six hours or let it set overnight. At the end of this time, check the consistency of the yogurt by unwrapping the towel carefully and tilting gently. If it is solid enough for your liking, refrigerate immediately. Serve only when thoroughly chilled.

This method of yogurt-making is very simple, but not as foolproof as the others, since it is difficult to maintain a constant temperature in the milk mixture. The warmer the spot where the mixture sets, the quicker it will thicken and be ready, providing the area is not over 115° F. If you're not successful using this method the first time, try it again until you find just the right spot in your kitchen and until you know how long to let it set.

YOGURT IN A THERMOS

An extremely simple way of making yogurt is to use a thermos bottle. A thermos is an excellent heat retainer. Once the starter has been stirred into the lukewarm milk, pour it into a wide-mouthed thermos, put on the lid and let it set four to six hours before refrigerating. This is practically foolproof, since the temperature is controlled for you.

Yogurt-making is a simple procedure. Once you become familiar with the method you have chosen you'll be able to make perfect yogurt every time.

Properly made yogurt should be rich and custard-like and have a creamy, slightly tart taste. Homemade yogurt will be sweeter than any store-bought variety. If, after refrigeration, there is a little water (whey) on top of the yogurt, don't worry, you haven't done anything wrong. This is natural, especially after it has set in the refrigerator for a few days. Open a commercial yogurt and you'll find water on top, too. Either mix it in or pour it off, but save it for using instead of water in other recipes, since the whey is high in vitamin B_{12} and minerals.

WHAT CAN GO WRONG

If you have trouble making yogurt the first time, check for the following problems:

1. Perhaps the milk mixture was disturbed while incubating. Even a few tilts or knocks can cause the whey to separate from the curd (as it does in cottage cheese). Instead of being a thick and smooth yogurt, it may be watery and lumpy and resemble cottage cheese.
2. Perhaps your mixture was too hot or too cool. If the mixture is too cool, the growth of bacteria will be retarded. If it's too hot, the bacteria may be killed. Add more starter, incubate longer and adjust temperature to correct.
3. Perhaps the milk or yogurt starter was not too fresh. The older either is, the longer it will take to incubate and the more starter you should use. For best results, neither should be more than five days old.
4. Perhaps you used a pure yogurt culture which might take longer to thicken than prepared yogurt.

Yogurt will keep well for about eight days under refrigeration if it is kept in an air-tight container. It can also be frozen for several months. Make sure that you save some to start your next batch. The new starter should be used before it is five days old.

YOGURT CREAM CHEESE

This simple-to-make yogurt product can be used just like regular cream cheese, on crackers for hors d'oeuvres, for sandwiches, and in cookies, pies, and other pastries.

To make yogurt cream cheese, make your yogurt by following one set of instructions discussed in this section. Instead of refrigerating the yogurt once it has formed, pour it into a colander lined with a triple thickness of cheesecloth. Catch the whey by placing a bowl under the colander. Allow the whey to drip for one minute, then lift up the four corners of the cheesecloth and tie them together. Hang the cheesecloth bag over the sink by suspending the bag from the faucet. Let the cream cheese drip for 6 to 8 hours, then remove it from the bag and store in the refrigerator.

RECIPES

Yogurt Cream Cheese Pie

½ pound yogurt cream cheese
⅔ cup yogurt
2 tablespoons honey
1 teaspoon pure vanilla extract
1 cup coconut

Blend together the cream cheese, yogurt, honey, and vanilla. Fold in all but 1 tablespoon of the coconut. Pour cream cheese mixture into a coconut or chopped nut pie shell and garnish with remaining coconut.

Yougurt-Fruit Freeze

1 cup yogurt
1 tablespoon lemon juice
½ cup pineapple or orange juice (unsweetened)
¼ cup dried and stewed apricots or peaches
1 tablespoon honey
2 egg whites

Mix the yogurt, lemon juice, orange or pineapple juice, stewed fruit, and honey together until all ingredients are blended. Put the mixture into a freezer tray and freeze until firm. Take it out, put into a bowl, and stir until mixture is smooth and creamy.

Beat the egg whites until stiff and fold them into the fruit mixture. Return to freezer tray and freeze. Yield: 3 – 4 servings.

Yogurt Cheese Cake

Crust:
1 cup breadcrumbs (dry)
1 tablespoon honey
4 tablespoons oil (corn or soy)
½ teaspoon cinnamon

Combine ingredients and pat into spring-form pan, covering bottom of pan and bringing up sides about 1½ inches.

Filling:
¼ cup cornstarch
1 cup yogurt
1 pound cottage cheese (low fat variety if possible)
rind of one lemon, grated
1 tablespoon lemon juice
¼ teaspoon salt
½ cup honey
2 teaspoons vanilla
4 eggs, separated

Using a blender, blend cornstarch and yogurt to dissolve cornstarch. Then add cottage cheese and blend until smooth. Add lemon rind, juice, salt, honey, vanilla. Blend to combine.

Beat egg yolks in mixing bowl until thick and lemon colored. Add cheese mixture and mix well.

Beat egg whites until stiff, but not dry. Fold into cheese mixture and pour into prepared pans. Bake at 300° F. for 50 minutes or until center is firm. When finished, turn off heat and with oven door closed, let cake cool in oven for 1 hour or more. When cool, loosen sides of spring pan and remove cake.

Note: Cheesecake may be topped with a cornstarch-thickened fruit such as unsweetened canned pineapple, or fresh strawberries, or blueberries. Yield: 12 servings.

Fruit Dressing

1 cup yogurt
2 tablespoons orange juice
1 teaspoon lemon juice
1 teaspoon orange rind
½ teaspoon lemon rind
1 tablespoon honey
¼ teaspoon grated or ground nutmeg
⅛ teaspoon ground mace

Blend all ingredients together in order given. Pour into a container with cover and place in refrigerator.

Serve with fruit salads or gelatin salads. Yield: 1¼ cups.

Carrot-Coconut Salad

2 cups shredded raw carrots
½ cup flaked or coarsely shredded coconut
½ cup pitted dates, coarsely chopped
2 tablespoons fresh lemon juice
¼ cup mayonnaise
¼ cup yogurt
1 tablespoon honey (optional)
⅛ teaspoon salt

Scrub carrots and shred. In medium bowl combine carrots, coconut and chopped dates. Mix mayonnaise, yogurt, lemon juice, honey and salt in a small bowl; blend well.

Pour dressing over salad and blend ingredients together. Chill before serving.

Line serving dish with greens; spoon salad in center and serve. Yield: 5 servings.
Note: Raisins may be substituted for dates.

Russian Dressing

1 cup yogurt
2 cups mayonnaise
2 cups catsup
4 teaspoons horseradish

Combine yogurt and mayonnaise with wire whisk. Stir in catsup and horseradish.

Store in covered glass jar in refrigerator. Yield: about 5 cups.

Green Onion Dressing

¾ cup thinly sliced green onions (including tops)
1 cup yogurt
2½ cups mayonnaise
2 tablespoons lemon juice
½ teaspoon salt

Combine all ingredients. Store in glass container in refrigerator, covered. Yield: about 4 cups.

Minted Cucumber Salad

1 teaspoon dried mint
½ teaspoon garlic powder
1 cup yogurt
1 large cucumber, peeled and sliced thin dark green lettuce leaves
4 red radishes, chopped

Mix mint and garlic powder into yogurt. Distribute cucumber slices on crisp leaves of lettuce and pour yogurt on top. Garnish with chopped radishes. Serve cold. Yield: 2 – 3 servings.

Cucumber Salad

2 – 3 cucumbers
salt
pepper
3 tablespoons yogurt
dill
parsley

Slice young cucumbers very thin, place in bowl and add sea salt and pepper to taste. Add three tablespoons of plain yogurt and some chopped dill. Garnish with sprigs of parsley after it has set an hour or more, then serve. If you prefer, use sour cream in place of yogurt. Yield: 4 servings.

Cooked Cole Slaw Dressing

3 tablespoons honey
1 cup homemade yogurt
1 teaspoon sea salt
2 eggs beaten
½ cup tarragon vinegar
1 teaspoon chopped celery

Cook in top of double boiler until it has become a smooth custard. Yield: 2 cups.

Honey-Yogurt Fruit Salad Dressing

½ cup yogurt
1 tablespoon lemon juice
½ cup honey
cinnamon, nutmeg or mace

Blend all together well, add spices and use on fruit salad.

Cooked Fruit Salad Dressing

½ cup apricot juice or any desired unsweetened fruit juice
4 tablespoons honey
2 beaten egg yolks
1 cup yogurt
cinnamon, or mace to taste

Use double boiler in which you mix juice, sea salt and honey. Add egg yolks gradually and beat well. Stir constantly until thickened. Remove from heat and let cool. When cool fold in yogurt and desired spicc. Yield: 1½ cups.

Beets with Yogurt

4 – 6 beets
bay leaf
parsley
basil
¼ cup yogurt
1 teaspoon honey

Slice beets very thin or grate, place in steamer over boiling water, add the beet tops along with a bay leaf, a sprig of parsley and one of basil. Let steam over boiling water for about 5 minutes. Remove the bay leaf and place beets in dish. Mix yogurt with honey, pour over beets and serve.

Stroganoff

2 medium onions, peeled and sliced
¼ cup oil
1½ pounds round steak, cut into 1-inch strips
½ teaspoon salt
few grains pepper
1 tablespoon soy sauce
1½ cups fresh sliced mushrooms
3 tablespoons water
2 tablespoons flour
1 cup yogurt
sprinkle of nutmeg

Cook onions in oil until tender, stirring occasionally. Add round steak and brown lightly on both sides. Add mushrooms, water, salt, pepper, and soy sauce. Cover and simmer slowly until tender (about 5 minutes). During that time mix together flour and yogurt. Remove round steak and onions from skillet when done. Take ¼ cup mushroom sauce from skillet and blend smoothly with yogurt and flour. Pour this mixture gradually into skillet, mixing well. Heat over low heat stirring constantly until thickened. Add round steak and onions just long enough to heat. Yield: 6 servings.

MAKING SEMI-HARD CHEESE

Cheese is one of those foods which has a certain mystique associated with it—perhaps because there are so many different kinds. Incorporated in this mystique is the idea that making cheese is something not to be attempted by amateurs. This just is not so. There are many amateur cheese-makers around who make some pretty good-tasting cheeses.

The first essential ingredient in cheese is lots of milk—the more the better. It does not pay to make a cheese smaller than three to five pounds as smaller cheeses dry out too much during aging. The second essential ingredient is a decent set of directions. And these are hard to come by. Current United States Department of Agriculture publications offer no practical information on making cheese. The only literature on cheese-making that is readily available is intended for commercial operations. Cookbooks written when large-scale food processing was still a thing of the future, provide instructions for making cheese, but the methods are too antiquated to be practical today. The best way to get solid information on cheese is to go to someone who makes a lot of it and ask him or her how it's done. This is just what we did. David Page is an organic gardener and an assistant professor of biochemistry at Bates College in Lewiston, Maine. He is also an amateur cheese-maker who turns out some great-tasting cheeses. Here are his directions for making what he calls a "non-descript, semi-hard, mild-to-sharp, pleasant-tasting, unadulterated cheese."

RIPENING THE MILK

Your cheese is doomed to failure (as I found out the hard way) if your milk is not ripened properly. The idea is to inoculate the milk to be used with a lactabacillus culture in order that other strains of bacteria don't get a chance to grow during aging and thus ruin the cheese. Obtaining a culture is simple—just use cultured buttermilk from the store. Take a quart of fresh cultured buttermilk and add about one-quarter cup to each of four clean, scalded quart Mason jars. Fill up the jars with fresh pasteurized milk. Seal with clean caps, shake, and allow to stand for twenty-four hours at room temperature. This gives you four quarts of fresh buttermilk, which you can also use for other things. You can put your culture in the refrigerator and use as needed. The culture can be perpetuated merely by repeating the foregoing process with your last jar of buttermilk at any given time. This is known as better living through bacteriology.

Get the biggest canning kettle you can find. I use a thirty-six-quart kettle which is pretty heavy when it is full. A twenty-four-quart canning kettle is all right, but your cheese will be smaller. The kettle must be enamelled. Stainless steel is fine but aluminum is a "no-no". Add the contents of one Mason jar of buttermilk to whichever kettle you are using. Then pour in sufficient milk to fill your kettle to within two inches of the top, and mix with a clean spoon. We use raw milk for making cheese, but pasteurized milk is OK too.

The milk should not be more than two days old (having been refrigerated of course) and may be skimmed if desired. Warm the milk up to 86 to 90° F. over low heat and allow to ripen for one to two hours. That is, set it aside; go and do something; when you're done—it's done.

FORMING THE CURD

Rennet tablets are supposed to be available at the corner drug store. Now that such stores have concentrated on such salubrious items as power tools and school supplies, mundane things like rennet are seldom to be found. If you can't get rennet locally, try one of the sources listed at the end of this section.

Dissolve one-quarter of a rennet tablet for each two gallons of milk in about one-half cup of cool water. (A bit too much rennet is better than a bit too little.) Make sure the ripened

1. Begin your cheese-making project by preparing the activator. Pour ¼-cup of cultured buttermilk into each quart jar. Fill the jars with milk and let them sit at room temperature for 24 hours.

2. Take one of these quart jars, whose contents has now turned into buttermilk, and pour it into a large pot of milk. Heat this milk to 86 to 90°F. and let it ripen for 1 to 2 hours.

3. In the meantime, dissolve ¼ of a rennet tablet in ½-cup of cool water for every 2 gallons of milk you have in your pot. Add this rennet solution to the warm, ripened milk. Cover the milk and let it stand for 1 hour.

milk is at about 86 – 90° F. (Check its temperature with a dairy thermometer). Add the rennet solution to the milk with plenty of stirring. Cover and let stand for about one hour. The way to tell if it's done is to stick your finger (washed of course) into the curd at an angle and lift up slowly. If the curd makes a clean break over your finger, it's done. If it's still the consistency of tired yogurt—be patient.

CUTTING THE CURD

When the milk has curdled, get a long knife and slice the curd up into cubes about one-half inch square— don't worry about getting nice looking cubes. Holding the knife vertically, slice in a parallel fashion in lines about one-half inch apart in one direction and then slice in the same manner in a direction perpendicular to the original slicings. Then slice at a sharp angle across one way following the original lines as best you can in order to undercut the curd and make into cubes. Do this undercutting in several directions until you think you have the curd pretty well cut up.

HEATING THE CURD

Place the whole works on a very low fire. (I use the low setting on my electric stove for eight gallons; smaller amounts require a lower setting.) Stir with a clean spoon constantly in order not to burn the curd at the bottom. As you stir with one hand, have your knife ready in the other ready to hack up any large pieces of curd that escaped the blade previously. Besides, hacking the curd helps to relieve the tedium of standing over the pot and stirring. As you stir, the contents will warm up—slowly. It should take about forty-five minutes to one hour to get from 86° F. to 105° F. You will note that the big lumps will get smaller and a yellowish cloudy fluid called whey separates. Keep the curds in motion during heating. When you've reached 105°F., take the pot off the stove and let stand for one hour. That should allow sufficient time for the curd to harden. What you will have is something resembling a white rubber bath mat at the bottom of a pot full of whey. Pour or scoop off the whey. (Save it to feed to chickens or pigs if you have them, or dump it on the garden or compost pile.) Leave the curds in the pot– it is easier to salt them if you do. For eight gallons worth of curds I add about three tablespoons of sea salt. The amount you add is purely a matter of taste. Using a spatula (pancake turner) slice into the matted curd and mix in the salt. At this point, the curd is quite well held together. A commercial cheese-making establishment would have a machine to chop up the curd–something like a compost shredder; but save yours for better things. Some whey will be sloshing around with the curds which will help to disperse the salt throughout the curds.

PRESSING THE CURDS

Go to the hardware store and buy ten or twenty yards of cheesecloth. The stuff you seem to be able to get around here is pretty cheesy so you must use a double thickness of cloth about a yard square. Place the double layer of cloth over a clean pail or bowl which is at least the same volume as the curd you have. Dump the curd (the spatula really comes in handy here) into the cheese cloth and strain out the remaining whey. After all the curd is in the cheesecloth, pick up the corners and form the curd into a ball by twisting the cheesecloth and squeezing the curd in the appropriate places. Hang up and let drain for about fifteen minutes. Now you should have a ball of curd of varying size depending on how much milk you started with. Fold a clean dish towel which is about two feet long into a multi–layered band about four inches wide by the length of the towel. Wrap this band around the side of the ball of curd as tightly as you can and fasten the end of the band with safety pins. What you have now is a ball of curd in a cloth girdle. Place the ball on a towel which is laid in the bottom of an eight-inch shallow bowl or deep plate. This will form the shape of the bottom of your wheel of cheese. Where the twisty part of the original cheesecloth is on the top of the ball, re-

4. When curds have formed, cut them into cubes and heat them over low heat until they reach 105°F. Stir constantly and have a knife at hand to cut up any large curds that should surface. After about 45 minutes the curds should resemble cottage cheese, and the cloudy whey should have separated to the top.

5. Remove the curds from the heat and allow them to harden for 1 hour. Then scoop off the whey and salt the curds.

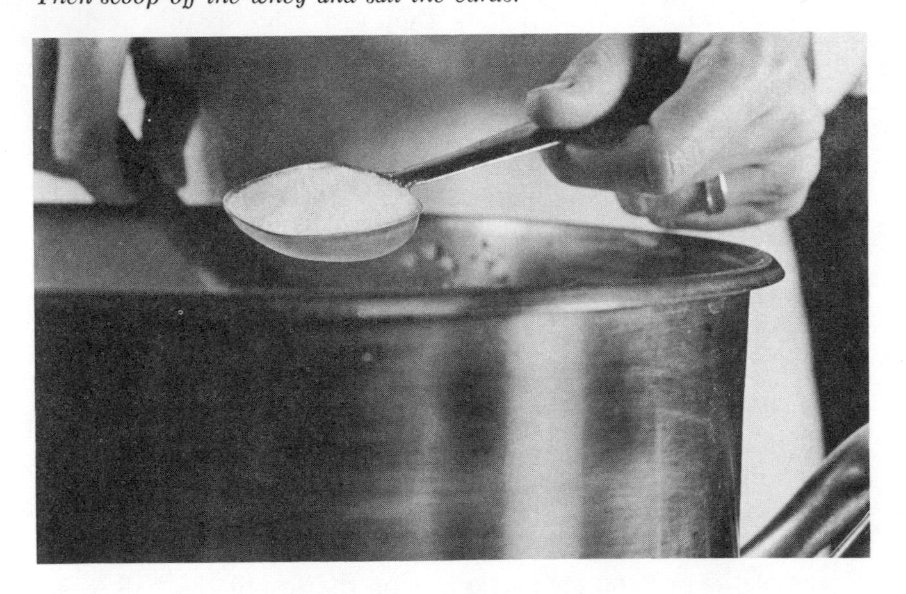

6. Strain the curds through cheesecloth, and when most of the whey has already drained off, pick up the 4 corners of the cheesecloth and squeeze the curds into a ball.

arrange the cloth so that when the top is compressed, the folds of the cheesecloth do not unduly indent the final wheel of cheese. Place a similar bowl over the top of the ball of curd and pile about forty to sixty pounds worth of books on top of everything and let stand overnight. Smaller cheeses need less weight. The idea is to press the individual curd granules into a solid wheel of cheese. There are various appliances one can make to do the pressing job, but I haven't had the time to make one. Using the erudite method given above, you have to watch out for several things. First, if your bowls are too deep, they will just come together under pressure and the curd won't get pressed. Also, if the books slip off at an angle, your final wheel of cheese will be lopsided—so be sure to distribute your weights above the curd evenly. A bit of intelligence and ingenuity could come up with a much better way of pressing the curd. After twelve hours, you will have a nice compact wheel of cheese. Strip off the cheesecloth and wind a fresh band of cloth around

7. After the curds have been pressed and then aged for about 6 days, rewrap the cheese in fresh cheesecloth and coat it with melted paraffin. Store the cheese for at least 60 days before eating.

where the old girdle was. Fasten again with safety pins. The reason for doing this is that the cheese at this point is quite plastic and if the sides of the wheel are not supported, the cheese will tend to flatten as it forms a rind.

Turn the cheese several times a day for five or six days, or until a good even rind has formed over the surface. Some directions say to dry for one to two days—don't believe 'em! Some mold may form on the outside, which is of no consequence.

AGING

By this time you are probably wondering if you will ever get to taste your fledgling cheese. Well, you must wait for at least sixty days. Wipe off the outside of the wheel of cheese and wrap tightly with one or two layers of cheesecloth. Heat one to two pounds of paraffin in a pot until it is good and hot, and brush the hot wax over the cheese or dunk portions of the wheel into the wax. Be careful of the hot wax. It is important to paraffin the entire cheese or it will dry out. The cheesecloth helps to keep the paraffin coating from cracking. Write the date on a piece of paper and glue it to the outside of the cheese with hot paraffin so you will know when the cheese is ready to eat. Place the finished cheese in a cool place on a clean surface and turn once every few days. If there is mold growth under the paraffin don't worry, it won't invade the cheese. If the cheese starts to swell as my first ones did, you got troubles. Such behavior indicates that your milk was not properly ripened and that a "bad" micro-organism is enjoying your cheese. Other than bad bacteria, the other big enemy of ripening cheese is small animals—most notably. At our place, an army of six cats (unfortunately all ours) gobbled down fifteen pounds of cheese in two days of feline gastronomic ecstasy. If you have no cats then worry about mice and rats eating your cheeses. If your wheels survive, they will taste pretty good after sixty to ninety days. The cheese will become sharper the older it gets.

This completes the basics of making a good home-made cheese. One can add various herbs and natural coloring agents to the milk after it has ripened in order to obtain a "different" cheese.

Sources of Rennet

Dairy Laboratories
2300 Locust Street
Philadelphia, PA 19103

Marschall Dairy Laboratory
14 Proudfit Street
Madison, WI. 53703

Chr. Hansen's Laboratory
9015 West Maple Street
Milwaukee, WI 53214

STORING THE CHEESE

Semi-hard cheese, like the one David Page makes, and others, like cheddar, American, Swiss, and Provolone, should be kept under refrigeration after aging. To prevent these cheeses from drying out, they may be wrapped in a thin piece of cloth or cheesecloth which has been moistened with water, a weak salt water solution, or a mild vinegar. Check the cheese occasionally and dampen the cloth when it becomes dry. To prevent the cut edges of a round of cheese from drying out, spread butter or a very light-flavored oil (sesame seed or safflower) on the cut ends of the round before wrapping and refrigerating it.

If you plan to keep the cheese for more than a few weeks, it may be frozen. Semi-hard cheeses do not freeze very well—they become a bit crumbly and begin to lose their flavor after a few months at freezer temperatures—but some people do not mind this slight loss in quality. If you wish to freeze your cheese, cut it into small pieces of about half a pound each, and wrap each piece well in moisture-proof paper. In small pieces, the cheese will freeze quickly, and there will be little damage done to the cheese's flavor and texture. Wrapped properly, semi-hard cheeses will keep for six months in the freezer.

In order to enjoy the full flavor of the cheese, it should be brought to room temperature before serving. Avoid exposing the cheese to high temperatures which cause the cheese to sweat and lose some of the fat captured in the curd.

STORING EGGS

Because each egg is intended by nature to house an unborn chicken, nature packages each one in its own protective shell. The shell is porous enough to permit oxygen and other gases to flow in and out through its walls, but its outer coating or membrane prevents bacteria and molds to enter which would otherwise contaminate the egg.

Alone, the shell will protect the eggs for a short time, providing it is kept cool. Brush, don't wash, dirt off eggs before you store them. People who vigorously wash off the dirt are also washing off the egg's protective membrane. If possible, store your eggs in a covered container to keep out objectionable odors that travel with gases through the shell's pores.

Eggs will keep at refrigerator temperatures for a week or two, but after that time their freshness fades. Both the white and the yolk begin to lose their firmness and become watery and runny. The yolk of an old egg will usually break into the white when the shell is cracked open, making separating the yolk from the white of old eggs a difficult, if not impossible, task.

OLD-FASHIONED METHODS

Before farmers had access to freezers, they devised some simple (but not always successful) means of preserving their excess eggs. Some farmers relied solely on the use of salt to keep their eggs from rotting. After gathering their eggs, they packed them in a large barrel or crock with plenty of salt and stored them in a cellar or spring house to keep them cool.

The majority, however, found some way to clog up the pores of the egg shells so that moisture would not escape and air could not enter. Eggs were rubbed with grease, zinc, or boric ointment, or submerged in a solution of lime, salt, cream of tartar, and water.

Probably the most popular way to seal egg shells was to waterglass them. By this method a chemical, sodium silicate, was mixed with water and poured in a crock which was filled with eggs that were about twelve hours old. The sodium silicate (which is used today to seal concrete floors and as an adhesive in the paper industry) would clog the pores in the shells and make them airtight.

Some people, even today, use waterglassing as a means of preserving eggs, but this storage method has its drawbacks. Eggs preserved this way are not good for boiling because their shells become very soft in the waterglass solution. The whites will not become stiff and form peaks, no matter how long they are beaten. No souffles, egg nogs, or meringues with waterglassed eggs. There is also a very good possibility that by consuming eggs stored in waterglass you would be consuming some of the undesirable chemical, sodium silicate. If you keep roosters with your hens, (which you'll do if you want to maintain a natural, happy environment for your hens and produce wholesome eggs for your family), waterglassing may not be a successful means of preservation for you. The life factor in fertilized eggs makes these eggs deteriorate more quickly than sterile, unfertilized eggs, and waterglassing may not be enough of a preventative against spoilage.

FREEZING

Freezing is the only way to keep eggs safely at home for more than two weeks. Eggs, both fertile and unfertile, will keep as long as six months in the freezer, if you prepare and pack them properly. The rule for selecting the right food for freezing applies for eggs just as it does for fruits and vegetables: choose only the very freshest. Eggs even a day or two old should be stored in the refrigerator and used within a relatively short time, as recipes call for them. Freeze only just-gathered eggs.

Eggs in their shell expand under freezing temperatures and split open. For this reason, they must be shelled and stored in appropriate containers. If you are storing eggs in rigid containers, leave a little headspace for expansion. You can separate the white from the yolk and freeze each separately, or you can store the eggs whole.

If you are freezing egg whites alone, they can be frozen as is, in air-tight containers. For convenience, pack as many eggs together as you will need for your favorite recipes. You can then thaw and use a whole container of egg whites at one time.

If you are packing yolks separately or are packing whole eggs, you will need to stabilize the yolks so that they won't become hard and pasty after thawing. To do this, add one teaspoon of salt or one teaspoon of honey to each cup of yolks. Twelve yolks make up one cup. Break up the yolks and stir in the salt or honey. Of course, it is necessary to mark on the container whether salt or honey was used as the stabilizer so that you won't ruin recipes by adding more salt or honey than you had intended.

If you are packing your eggs whole, you will also need to stabilize them with salt or honey. Add one teaspoon of salt or honey to each cup of whole eggs. There are about five whole eggs in one cup. Scramble the eggs with the salt or honey before packing and freezing. Whole eggs can be packed together in one container or they can be packed individually by using a plastic ice cube tray. To pack eggs separately, measure three tablespoons of whole scrambled eggs (which equals one whole egg) into each separate compartment of the ice cube tray. Place the filled tray in the freezer, and when the eggs have frozen, pop them out and store all the egg cubes in a plastic bag. By so doing you will be able to take from the bag and thaw just as many eggs as you need at one time.

Eggs should be thawed completely before using. They thaw at refrigerator temperatures in about nine hours and at room temperatures in about four hours. If frozen properly, thawed eggs have the taste, texture and nutritional value of fresh eggs and can be used successfully in all recipes calling for eggs. To make up one egg from separately frozen whites and yolks, measure out one tablespoon of yolk and two tablespoons of white. Eggs should be used soon after they thaw, as they deteriorate rapidly.

MEATS

PREPARING BEEF VEAL LAMB AND PORK FOR STORAGE

Obviously, there's quite a bit of work involved in getting meat on the hoof slaughtered, dressed, and cut before it can be wrapped and frozen, canned, or cured and smoked. A butcher can do all this preliminary work for you, or, if you have the proper equipment and the knowhow, you can do the whole job yourself.

DOING YOUR OWN BUTCHERING

If you're going to butcher yourself, you have a lot of things to consider. First, do you have a place in which to butcher? It must be clean, cool, dry, and well-ventilated, with a ready source of water and a stove for heating water. A garage, basement, or outbuilding can be converted for this purpose, providing there is enough head room to hoist the carcass. Many farmers do their slaughtering and dressing outdoors and then bring the carcass inside to cut. If they do work outdoors, they usually choose a dry, cool fall day on which to do the job. If you are working indoors and the temperature of the room you are using is higher than 38° F., a large cooler will be needed in which to chill the meat, and in the case of beef and lamb, also to age it.

Equipment

The equipment to have on hand for butchering includes something to hoist and suspend the carcass, like strong hooks, a brace extending from an outbuilding or tree (if you're working outdoors), or a heavy rod suspended from the ceiling. A block and tackle, windlass, or chain hoist would be very helpful in hoisting the carcass, especially if you're butchering steer. Rope, buckets, a thermometer to measure the temperature of water, a meat thermometer, a stunning instrument, cleaver, meat saw, hog scraper, hand hooks, whetstone, steel, and a good set of sharp knives are necessary. You may want to have access to a meat grinder, also. If you're butchering hogs, you'll need a scalding vat or watertight barrel. For slaughtering lambs, a low bench or box is necessary.

Slaughtering and dressing animals is an exacting job, and space doesn't allow us to explain the specifics of butchering in this book. For good, detailed information, we refer you to the booklets published by the U.S. Department of Agriculture on the subject. These booklets, which contain lists of equipment, diagrams, and easy-to-follow directions, can be obtained for a small charge from your state or county agricultural station or the Superintendent of Documents at the U.S. Government Printing Office in Washington, D.C. There are also some good books on butchering meat that you might want to consult. You'll find a list of these books at the end of this chapter.

Chilling Meat

After the animal has been slaughtered and dressed, the carcass must be chilled promptly to rid it of its aminal heat. If not chilled, the meat will spoil more rapidly because destructive bacteria thrive at normal body temperatures. Chilling also makes the meat easier to cut. To hasten chilling, cut off excess fat in the crotch and split the carcass. Hang the carcass to chill in a well-ventilated, cold area with a temperature between 32 and 40° F. A lamb or veal carcass kept in an area with a temperature between 32 and 40° F. should be chilled to 40° F. within twenty-four hours. A beef carcass may require forty or more hours to chill to this temperature. A meat thermometer inserted into the thickest part of the carcass will show you when the meat is properly chilled.

Slaughtering on the farm is best done on a fall or early spring afternoon so that the carcass can begin to cool down during the night when temperatures just above freezing prevail. (If you will be curing and smoking your meat it is better to slaughter in the fall. Cold winter temperatures help to retard growth of bacteria responsible for spoilage.) If night temperatures should rise above 40° F., cut the carcass in half and then into a few big pieces and immerse them in clean barrels filled with water, ice, and about three pounds of common salt. This solution, which is colder than ice water, but warmer than solid ice, will help chill meat, but will not bring it down to freezing temperatures.

Never try to chill a carcass quickly by exposing it to freezing temperatures or by packing it in solid ice. A carcass should never freeze because the thin layer of ice that forms on the meat surface will prevent the proper escape of animal heat from the center of the meat. If a carcass should freeze, thaw it slowly at temperatures no greater than 40° F.

Aging Beef and Lamb

Beef and lamb are aged after chilling. This means holding the meat at a temperature of from 32 to 38° F. for a few days to a few weeks to increase tenderness, and in some meats, to bring out the flavor. Aging is best done in a refrigerated area where the temperature does not vary a great deal. If a cooler is not available, aging may be done in a clean, cool and well-ventilated place, free from animals and insects.

The length of the time meat should be aged varies. Lamb can be held for one to three days, while older sheep (mutton) can be aged five to seven days. Beef with little external fat should not be held for more than five days after slaughter. Beef with a good amount of external fat can be aged five to eighteen days. Fat acts as a protection against bacteria in the air that would grow on the meat surface if it were left exposed.

People who plan to store much of their lamb or beef in the freezer for more than six months should limit the aging period. Experiments at Pennsylvania State University have shown that the length of the aging period has a direct bearing on the storage life of meat because it permits oxygen absorption by the exposed fat.* As we have mentioned previously, the more oxygen

*P. Thomas Ziegler, *The Meat We Eat,* Danville, Illinois: The Interstate Printers & Publishers, Inc., 1965.

absorbed by the fat, the quicker the rate of rancidity. Aged meat shows higher peroxide values and shorter storage life than forty-eight-hour chilled meat. In addition, the experiments at Penn State showed that aging does not influence the tenderness of meat that is frozen more than a month. Experimenters found that although aged meat is slightly more tender during the first month of storage, this advantage disappears in subsequent months, when the aged and the forty-eight-hour chilled meat are on a par for tenderness. Pork and veal should never be aged, but frozen, cooked, or cured as soon as the animal heat is gone.

Cutting the Meat

Whether you are cutting your own meat or having a butcher cut to order, you'll want to get those cuts that best suit your family's needs. Ask yourself these questions before the cutting begins: Do I want more steaks or chops than roasts, or would I prefer it the other way around and get more roasts? Do I want some meat left around the bones for meaty soup bones, or would I prefer it removed and ground? Do I want some meat cut into stewing chunks or have it ground for hamburger meat? You do

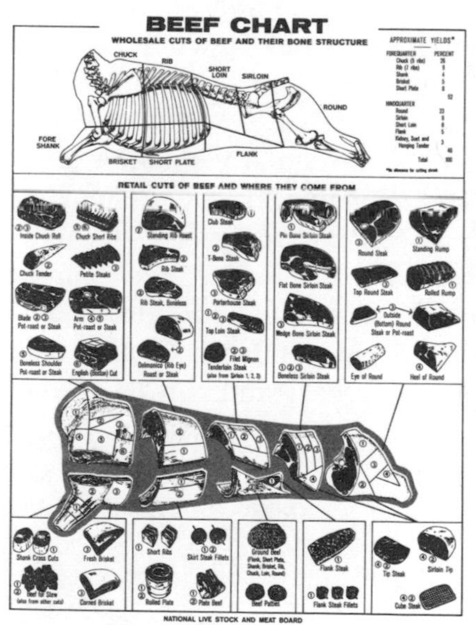

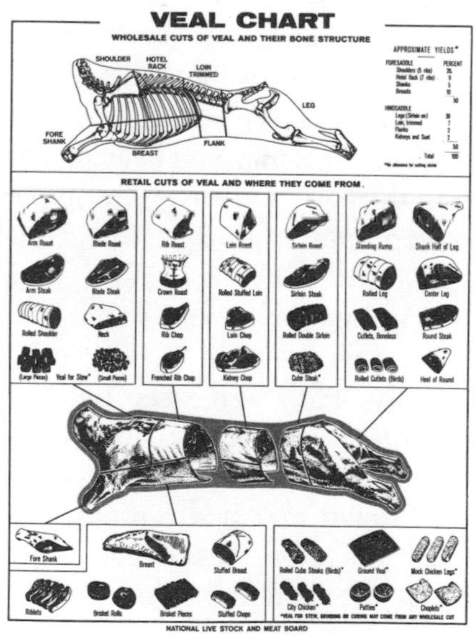

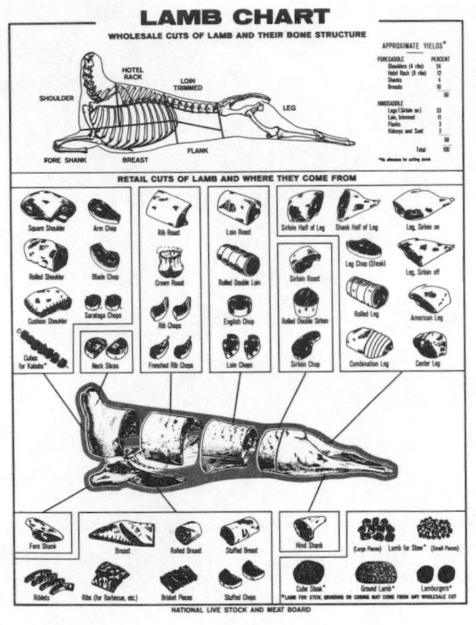

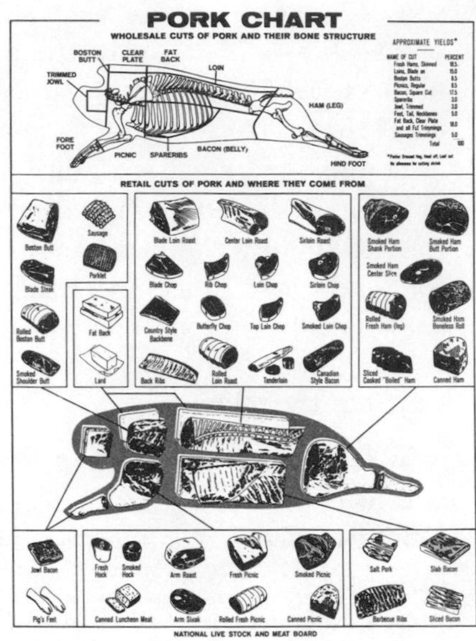

have these choices because there are several different ways certain parts of the carcass may be divided.

The size of the roasts, the thickness and number of steaks and chops, and the amount of ground beef and stewing beef are determined by the size and preferences of your family as well as your method of storage.

For freezing, you'll want your cuts as regularly square-shaped as possible for easy wrapping and packing. If you're smoking, keep your hams whole and your bacon in large pieces. Canning requires smaller pieces of well-trimmed meat. If you're canning your meat or freezer space is limited, you should consider boning as much meat as is practical. Boned cuts fit into jars and cans easily. They also wrap better for the freezer and can be packed tighter, with less chance of the wrap's tearing, when no large bones are protruding. Depending upon the cut, boned meat can be rolled and tied, ground for hamburger meat or sausages, cut for boneless steaks and easy-to-carve roasts, or prepared as stewing meat or bacon.

Approximate yields of trimmed beef cuts from animal having a live weight of 750 pounds and a carcass weight of 420 pounds

Trimmed cuts	Yield	Live weight	Carcass weight
	Pounds	Percent	Percent
Steaks and oven roasts	172	23	40
Pot roasts	83	11	20
Stew and ground meat	83	11	20
Total	338	45	80

Approximate yields of trimmed beef cuts from dressed forequarters weighing 218 pounds

Trimmed cuts	Yield	Weight of forequarters
	Pounds	Percent
Steaks and oven roasts	55	25
Pot roasts	70	32
Stew and ground meat	59	37
Total	184	84

Approximate yields of trimmed beef cuts from dressed hindquarters weighing 202 pounds

Trimmed cuts	Yield	Weight of hindquarters
	Pounds	Percent
Steaks and oven roasts	117	58
Stew, ground meat, and pot roasts	37	18
Total	154	76

Courtesy U.S. Department of Agriculture

Have the larger bones cracked for making stocks. You may wish to make very concentrated soup stock from the bones and freeze this broth instead of wrapping and freezing awkwardly shaped, clumsy bones that take up much space. If you are canning your meat, the only practical way to save the juices and gelatinous extractives from the bones is to make soup stock and can it.

Good information on various ways to cut beef, veal, lamb, and pork can be obtained from your state or local agricultural extension station or the Superintendent of Documents. This information can also be found in most of the books listed at the end of this chapter.

Approximate trimmed pork cuts from a hog having a live weight of 225 pounds and a carcass weight of 176 pounds

Trimmed cuts	Yield	Live weight	Carcass weight
	Pounds	Percent	Percent
Fresh hams, shoulders, bacon, jowls	90	40	50
Loins, ribs, sausage	34	15	20
Total	124	55	70
Lard, rendered	12	15	27

Yields of trimmed lamb cuts from a lamb having a live weight of 85 pounds and a carcass weight of 41 pounds

Trimmed cuts	Yield	Live weight	Carcass weight
	Pounds	Percent	Percent
Legs, chops, shoulders	31	37	75
Breast and stew	7	8	15
Total	38	45	90

Courtesy U.S. Department of Agriculture

IF YOU'RE HAVING YOUR MEAT BUTCHERED

Not too many people who raise a few meat animals a year for family use do their own butchering any more. Instead, they hire a butcher to do all or part of the work for them.

Judy and Arnold Voehringer, of Kempton, Pennsylvania, raise their own beef cattle organically, and until two years ago, also raised hogs. They haven't found it economical, on their small-scale operation, to butcher their meat themselves. A local butcher comes for the animals and takes them back to his shop in a truck to slaughter, dress, and cut them. The Voehringers take the cut meat home to wrap and freeze. Although wrapping the meat from a 900-pound dressed steer is an all-day job for one person, the Voehringers feel it is enough of a savings, money-wise, to justify all the work; they save 7 cents per pound by wrapping and quick-freezing the meat themselves. Besides, there are usually enough friends and family members around on wrapping days to make the job easier.

The Voehringers feel that it is very important to take their animals to a butcher with whom they can feel confident. For this reason, they prefer to deal with local butchers whom they can get to know personally. They've learned that small operations will give them—and their steer—plenty of attention.

This personal attention was especially important to the Voehringers when their first few steers were slaughtered. The butcher took time to explain the different cuts of meat to them and give them tips for wrapping and freezing at home. Judy stills helps the butcher when he is cutting the meat so that she can tell him how she wants the cuts: how thick to make the steaks, what size roasts she wants, how much fat to trim off, and so on. It is against the policy of most larger meat companies to permit customers in the cutting room. Rather, they cut the meat according to order sheets customers fill out when their animals are brought in for slaughter. Typical order sheets might look like these:

Beef

Name ....

Address ....

Cost of Hauling .... Slaughtering ....

Live Weight ....

Cutting Date .... Cut-By ....

Whole Carcass .... Wrapped By ....

Front .... Hind .... Side ....

Weight .... Price ....

Number of People in Family ....

Round

Rib

Roasts

Steaks

Soup Bones

Boiling Beef

Hamburger

Variety

Pick Up Date ....

Phone ....

We Wrap & Freeze

PORK

Name ....

Address ....

Date .... Hogs ....

Weight .... Live .... Dressed ....

Back Bone .... Pork Chops ....

Number of People in Family ....

Shoulders

Hams — Cure — We ....

They ....

Bacon

Knuckles

Stomachs

Sausage

Scrapple — Our Dishes ....

Their Dishes ....

Pudding — Dishes ....

Casings ....

Variety

Lbs. of Lard — Our Cans ....

Their Cans ....

Lbs. of Sausage

Beef for Sausage

Pick Up Date ....

Phone ....

We Wrap & Freeze

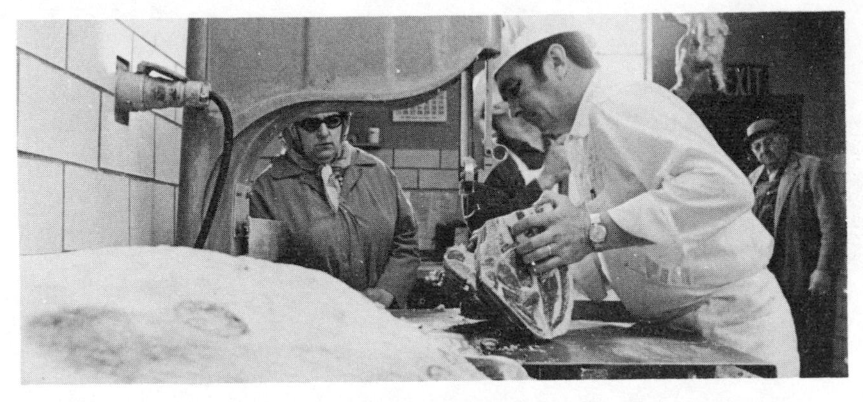

Most small meat companies that do private butchering invite their customers into the cutting room so that the customers can watch the butcher and tell him just how they want their meat cut.

Although filling out such an order form will allow you to choose pretty much how you want your meat cut, you have more of a say in the matter when you're standing next to the butcher during the whole process.

For practical purposes, most large meat operations collect the beef to be ground from all the steers butchered that day and put it together in a large meat grinder. What comes out is a mixture of meat from all the carcasses. The butcher makes note of the amount of ground meat belonging to each customer so that each gets his fair share, but certainly the ground meat that each customer gets is not all from his own steer. This doesn't matter to most farmers because the beef from one steer is almost the same as that from the next, but it does make a difference to the farmer who raises his steers organically. The Voehringers prefer working with small meat operations because there butchers grind the meat from each steer separately.

Making Headcheese, Liver Sausage, Scrapple, and Sausage

Early settlers could not afford to waste any part of their butchered hogs; the slaughtered animals were too valuable to them. They ate some of the pork fresh, cured the hams, shoulders, and bacons, and made sausage out of the lean meat scraps. The feet were pickled, the tails were used for stew, the fat was cut off and rendered for lard, and the heads, skins, organs, and bones were boiled to make dishes like headcheese, liver sausage, and scrapple.

Headcheese (or Souse)

Prepare the hog's heads for boiling by removing any hairs or bristles that remain. Cut out the eyes and quarter the heads. Let the cut-up heads soak in a pot of fresh water at least 7 hours to remove the remaining blood. Then rinse the head thoroughly in running water. Put the heads in a large heavy pot and add any other scraps you wish, like bones, hearts, and skins. Before putting the skins in the pot, put them in a sack made of a double thickness of cheesecloth so that they can be easily removed when they are tender. Skins are tender when they are easily pierced with your fingers. Cover the head and scraps with water and simmer until meat slips from the bones.

Remove the skins and grind them using a ⅛-inch-hole plate. Prick the meat from the bones and grind it with a ½-inch-hole-plate. Mix the ground meat and ground skin together with enough broth to make the mixture the consistency of a soft cake batter (the remaining broth may be saved to make liver sausage or scrapple). Return the mixture to the pot and add your spices. To each hog's head add the following:

1 tablespoon ground mixed spices (this may include garlic, savory, and onion powder, or sweet marjoram and ground cloves)
¼ teaspoon red pepper (optional)
2 tablespoons black pepper
1 crushed bay leaf
2 tablespoons salt

Bring the mixture to a boil and remove from heat. Pour into loaf pans and chill. This keeps well in the refrigerator and may be frozen for up to 2 months.

Liver Sausage

Liver sausage is a variation of headcheese. It is made by adding cooked pork livers to the cooked heads, tongues, skins, boned meat scraps, and broth. The livers should not constitute more than about 20 percent of the ingredients by weight.

To make liver sausage, cook 3 pounds of liver for 10 to 20 minutes, until done. Do not cook the liver more than 20 minutes because it will become crumbly. Add the cooked liver to 12 pounds of cooked meat scraps and grind the mixture moderately fine. Add 5 pounds of broth to make a soft, but not runny, mixture. Season to taste. The quantities below may be used as guidelines for seasonings:

½ cup salt
3 tablespoons black pepper
2 tablespoons ground sage (optional)
1 teaspoon red pepper (optional)
1 tablespoon allspice (optional)

The sausage may be poured into loaf pans and stored as headcheese, or it may be stuffed into beef casings and smoked to make the Braunschweiger type of liver sausage.

Scrapple

Scrapple or pann haas is a popular Pennsylvania Dutch breakfast dish made famous in Philadelphia. It is made by combining meat scraps with broth (which may be that remaining from the liver sausage or headcheese) and thickening it with corn meal or other cereal.

To make scrapple, cook heads, bones, or any other meat scraps and water in a heavy pot until meat falls easily from the bones. Remove the meat and grind finely. Strain the broth and return it to the pot with ground meat. Bring to a boil and slowly add the cereal thickener, stirring constantly to avoid lumps. Usually the cereal added is corn meal, but some of the corn meal may be replaced with buckwheat flour in a ratio of 2 parts corn meal to 1 part buckwheat flour. A small amount of wheat germ may also replace some of the corn meal.

To make a good-textured scrapple with a rich flavor, we recommend using the following proportions of ingredients:

8 pounds meat
6 pounds broth
2 pounds cereal

Boil the mixture for 30 minutes, stirring constantly to prevent scorching.

A few minutes before removing the mixture from the heat, season it with:

1 crushed bay leaf
1 tablespoon sage (if desired)
2 tablespoons salt
1 tablespoon sweet marjoram
2 tablespoons pepper
½ tablespoon ground nutmeg (if desired)
1 teaspoon red pepper (if desired)
2 teaspoons onion powder (if desired)

When the mixture has thickened and begins to leave the sides of the pot, pour it into loaf pans and chill quickly. Scrapple may be frozen for up to 2 months. To serve, slice and fry quickly, as you would sausage patties.

Sausage

For sausage, choose meat scraps with no skin, gristle, blood clots, or pieces of bone remaining. Good sausage meat should be about 25 percent fat. Too much fat makes a heavy, greasy sausage that shrinks a great deal during cooking, and too little fat makes a hard, dry sausage. If you are making a large quantity of sausage, it is a good idea to add seasonings sparingly. More spices can always be added if the meat is too mild, but none can be removed if the meat is too hot. Add your spices, mix them in well, and test the sausage meat by shaping some into a patty and cooking it. Taste the sausage and correct the seasonings. If the sausages are to be smoked, use a little less seasoning, as smoking will slightly dry out the meat and bring out the flavor of the spices. If you plan to freeze your fresh sausage meat, it is a good idea to add the seasoning after the meat has thawed, rather, than before you store it in the freezer. Spices shorten the freezer life of meats. Fresh sausage meat keeps safely at freezer temperatures for one to two months.

While searching through old recipes for sausage we discovered that just about every farmer who butchered hogs had his own special seasoning formula. We've chosen just three recipes from thousands in print.

Mild Sausage

10 pounds ground pork scraps
6 level tablespoons salt
4 teaspoons sage
4 teaspoons white pepper

Moderately Spicy Sausage

10 pounds ground pork scraps
5 tablespoons salt
2½ teaspoons dry mustard
5 teaspoons black pepper
2½ teaspoons ground cloves
5 teaspoons ground red pepper
6½ tablespoons ground sage

Spicy Sausage

10 pounds ground pork scraps
12½ tablespoons salt
10 teaspoons ground sage
10 teaspoons ground red pepper
2½ teaspoons ground allspice
5 teaspoons black pepper
2½ teaspoons honey
2½ teaspoons cayenne pepper
1 teaspoon powdered cloves
¼ teaspoon thyme

Mix the seasonings and then work them thoroughly through the meat. Sausage meat may be packed loose for freezer storage, or it may be stuffed into natural casings, cellulose, or muslin bags and stored in the freezer or smoked. For smoking sausage, see directions for smoking in the section on curing and smoking meats.

Baked Sausage Meat with Sweet Potatoes

4 large sweet potatoes
1 pound loose sausage meat
4 large apples
salt
honey
milk
butter

Boil the sweet potatoes until tender and cut into thin slices. Cover the bottom of an oiled baking dish with half the potato slices. Make 4 flat patties from the sausage meat and brown them lightly in a skillet. Drain the patties and place them over the potato slices. Slice the apples and place them over the patties. Sprinkle salt and honey to taste over the apples, and cover with the remaining potato slices. Brush the potatoes with milk and dot with butter. Bake for 45 minutes at 350° F. Yield: 4 servings.

Baked Stuffed Apples

6 large cooking apples
1 cup sausage meat
1 teaspoon salt
honey (optional)

Clean out the center of the apples by scooping out the cores and pulp (be careful not to cut through to the bottom of the apples). Chop the pulp and combine it with the sausage meat and salt. Sweeten with a little honey, if desired. Stuff the apples with this mixture, place upright in a baking dish, bake in a 375° F. oven until tender, about 45 minutes. Serve with brown rice or noodles. Yield: 6 servings.

Sausage With Gravy

1 pound pork sausage
1 tablespoon butter or drippings
1 cup bouillon
½ cup onion
½ cup water
½ teaspoon flour

Fry the sausage over low heat in drippings or butter until brown. Remove, then add sliced onion; add the flour. When brown add bouillon or soup stock. Cook a few minutes. Pour gravy over sausage and serve with sauerkraut. Yield: 2 – 3 servings.

Stuffed Pig's Stomach

1 fresh pig's stomach
1 pound smoked sausage
2 to 3 potatoes
salt and pepper
1 onion

Clean the pig's stomach thoroughly, removing the membranes and rinsing in warm water. Dice the raw potatoes and the onion. Cut the sausage into thin slices. Mix together the potatoes, onions, salt, and pepper and stuff the stomach with this mixture. Sew up the stomach and roast it in a 350° F. oven for 3 hours. Serve with applesauce. Yield: 4 servings.

Sausage and Sweet Potato Filling

2 large sweet potatoes
1 cup sausage meat
3 tablespoons chopped onion
1 cup diced celery
2 cups wholewheat bread-crumbs
3 tablespoons parsley
salt and pepper

Boil the potatoes, skin them, and mash. Brown the sausage meat in a skillet lightly and then remove from pan. Sauté the onion and celery until translucent. Mix the cooked onion, celery, and meat. Add the mashed potatoes, bread crumbs, and parsley, and season with salt and pepper, if necessary. Mix well. This filling should be sufficient for a 10-pound bird.

Potato Sausage Pie

6 or 8 potatoes
1 onion
1 pound loose sausage meat
1 teaspoon minced parsley
1 top pie crust

Slice the potatoes into rounds and dice the onion. Put the potatoes in salted boiling water and cook for a few minutes, until tender. Saute the meat and onion in a skillet. Drain the potatoes, saving 2 cups of water. Fill a casserole dish with alternating layers of potato slices and sausage meat. Add the water and sprinkle with parsley. Place a pie crust over the top and bake at 450° F. for 10 to 15 minutes, or until pie crust is golden. Yield: 4 servings.

Fried Sausages

2 pounds pork sausages
2 eggs
1 cup wholewheat bread-crumbs
½ cup flour
1 tablespoon butter or drippings

Salt the sausages, dip in a mixture of white of egg, flour, and bread crumbs. Fry slowly to a nice brown crisp color. Yield: 6 servings.

Peppers Stuffed with Scrapple

1 egg
½ pound Scrapple
1 onion chopped
1 cup fine wholewheat bread-crumbs
6 green peppers
½ teaspoon paprika
1 teaspoon chopped parsley

Fry the scrapple lightly in slices, in butter, and then mix the scrapple, the drippings, the slightly beaten egg, onion, paprika, parsley, and breadcrumbs into a thick paste. Remove the ends and stems of the peppers and stuff them with the mixture. Lay them on their sides in a bake pan containing a little water and bake for 30 minutes, turning occasionally. Yield: 4 servings.

Scrapple Croquettes

1 cup scrapple
2 eggs hard boiled
1 cup cooked brown rice or mashed potatoes
½ cup wholewheat bread-crumbs
1 teaspoon minced parsley
Salt, pepper
1 egg, beaten

Mix the scrapple, the rice or potatoes, and the hard-boiled eggs, chopped fine. Season with parsley, salt, and pepper. Shape into croquettes with beaten egg and bread crumbs, fry in deep fat. Yield: 3 – 4 servings.

Stuffed Cabbage

½ pound scrapple
1 onion, chopped
1 tablespoon arrowroot powder
1 tablespoon chopped parsley
1 cup fine wholewheat bread-crumbs
1 cup milk
1 egg, beaten

Put the scrapple in a bowl and mix with the other ingredients. Dress a well-formed head of cabbage and cut the top off neatly to make a "lid." Hollow out the center, to make a shell, with a hollow about 3 inches wide and 3 inches deep. Fill with the scrapple mixture and replace the "lid." Tie it in a cheesecloth and boil in salted water until tender. Such scrapple mixture as is left form into balls, dredge with flour and fry brown. Serve them on a platter around the cabbage. If desired make a tomato sauce to accompany it. Yield: 4 servings.

DRESSING POULTRY

Dressing poultry is a far less complicated operation than dressing meat, and is commonly done on the farm. No specific area is needed to slaughter and clean birds, unless you have a lot of chickens, turkeys, geese, or ducks to process. You can slaughter your birds as you need them or do many at a time for freezing or canning. With just a little improvisation, you can do the entire job outside, in a basement, garage, or outbuilding. Naturally, when dressing poultry, your prime concern should be to end up with a bird that is attractive and free from contamination. This means working carefully and quickly, with equipment that is clean and in good working condition. Your processing area, whatever it may be, should be a place that is clean and free from flies, with a ready source of water, and a stove for boiling water. It should also provide enough clean space where birds may be placed between processing steps.

EQUIPMENT

Equipment should be clean and ready to use before you choose your birds for slaughter. Like larger animals, birds also require a device to hoist and suspend them for dressing. There are instruments designed just for this purpose. A killing cone is like a funnel with the pointed end removed. The bird is placed inside this device, with its neck through the narrow opening. Killing cones come in various sizes, and it is important that you have one that fits the bird snugly. Shackles is a metal device that suspends birds by their feet. It may be used in place of the killing cone. If neither of these are available, you can suspend poultry by their feet with rope. Tie a short piece of rope to a convenient ceiling beam or support, ceiling hook or, if you're working outside, to a tree limb. Attach a two-inch square block of wood on the free end of this piece of rope so that you can make an adjustable loop to hold the bird's foot. Suspend the bird by inserting each foot in one of these loops.

For slaughtering and dressing you will need a number of different tools, including at least one boning knife, small rounded knives without cutting edges for scraping pinfeathers, shears for cleaning giblets, and a lung scraper to remove the lungs. Knives and shears should be kept sharp during the operation with a whetstone and steel.

You will also need a few containers. A watertight container that is big enough to hold birds for scalding is necessary. A clean, galvanized ten-or twenty-gallon garbage pail is ideal for the job. A container will be needed to hold feathers and inedible viscera until they can be disposed of. A paper-lined cardboard box or plastic-lined bucket, at least two-foot square, will work well. You will also need a container to hold the giblets, as they should be kept separate from the rest of the bird during processing. Any clean pot, pail, or bucket will be sufficient. A clean pail or bucket, big enough to hold dressed birds, ice and water will be needed for chilling. A work table, big enough to allow you to work freely, is also necessary. Pick one that is of a convenient height for you so that you can work comfortably, without having to stoop over unnecessarily. It should be sturdy, with a clean work surface.

Have on hand a durable thermometer that will register temperatures between 120 and 212° F. to measure the temperature of the water for scalding.

For information on the actual slaughtering and dressing of poultry, we refer you to the books and pamphlets listed at the end of this chapter.

PREPARING BIRDS FOR SLAUGHTER

Twenty-four to thirty-four hours before slaughter, pen those birds to be slaughtered and fast them. The cage should be clean, so that the bird's feathers will not get soiled. It should have a wire bottom so that birds cannot touch the ground to pick up feathers and litter. Fasting reduces the chance of contamination of the carcass because it cleans the digestive tract of feed and ingested matter. Birds should be given water during this fasting period, however so that they will not dehydrate. The skin of dehydrated birds is unattractive when the feathers are removed; it appears dark, dry and scaly.

CHILLING THE BIRD

As with beef, veal, lamb, and pork, poultry must be chilled after it is dressed to remove normal animal heat. Chilling reduces the temperature of the carcass enough to retard growth of bacteria that would otherwise lead to spoilage of the meat. Chilling also makes the carcass easier for cutting and handling. To pre–chill the carcass, put the bird in a stopped-up sink or in a clean watertight container filled with clean water that is safe for drinking. Allow this water to run slowly so that there is a constant overflow. If this is not possible, change the water periodically. This prechilling has two purposes: It helps to cool the bird, and it further cleans the animal. If you are dressing more than one bird at a time, add each to this water as it is cleaned.

Once the bird is sufficiently pre-chilled (it should be cooled down to water temperature), it must be chilled to 40° F. before further processing. This is done by placing the bird in a container filled with ice and water. Large capons will require three or more hours to chill to 40° F. Turkeys that are to be frozen should be held in 40° F. chill water for eighteen to twenty-four hours before wrapping and freezing. Do not let birds freeze during chilling by packing them in ice only or exposing them to freezing temperatures. If birds should freeze, thaw them slowly in water no warmer than 40° F. Sudden changes in temperature will lower the quality of the dressed poultry. Once the carcass has reached 40° F., remove it from the ice water, hang it by the wing and let it drain ten to thirty minutes before wrapping for freezing or refrigeration, or for canning.

CUTTING THE BIRD

If you are canning your poultry, you will want to cut up the bird. The breast should be split and then cut along each side of the backbone so that this bone can be removed. The breast may then be cut to fit your glass jars or tin cans. Thighs should be separated from the drumsticks. Thighs may be boned, if you wish. Wings should not be canned; there is not enough meat on them to make canning them worthwhile. Rather, add the wings to other bones and simmer in water for broth or soup stock. (For more information on canning poultry, see the section on canning meat.)

You should cut up your chicken if you are canning it or want to wrap it compactly to save freezer space. Before you begin, chill your bird in ice water and let it drain. Then follow the steps shown here for cutting up poultry.

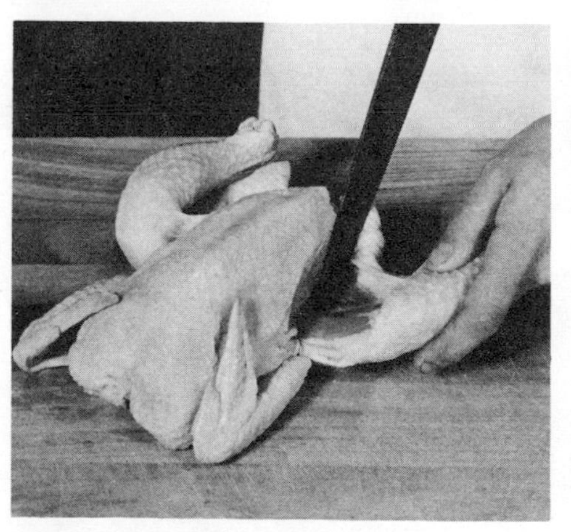

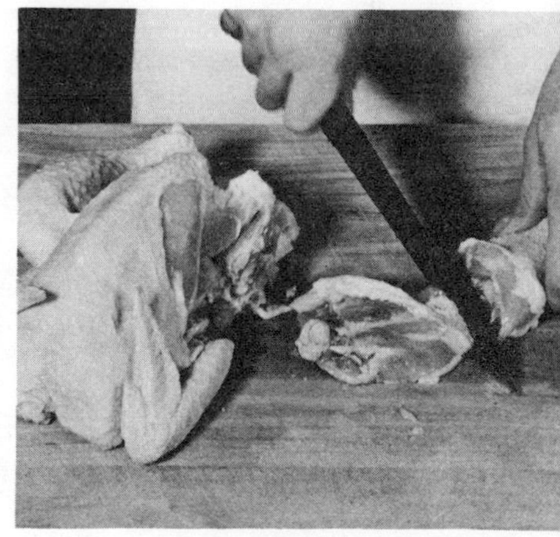

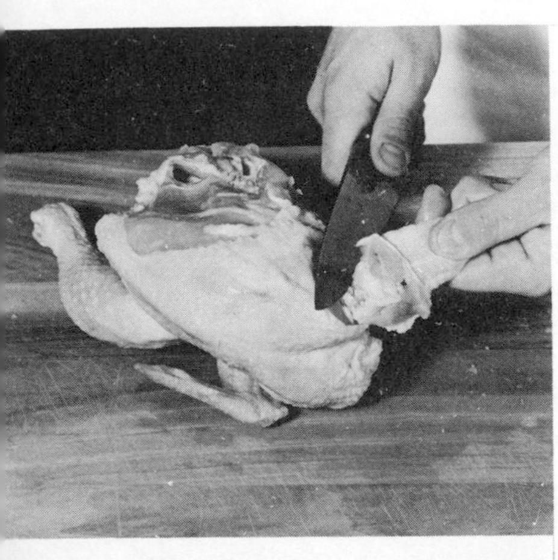

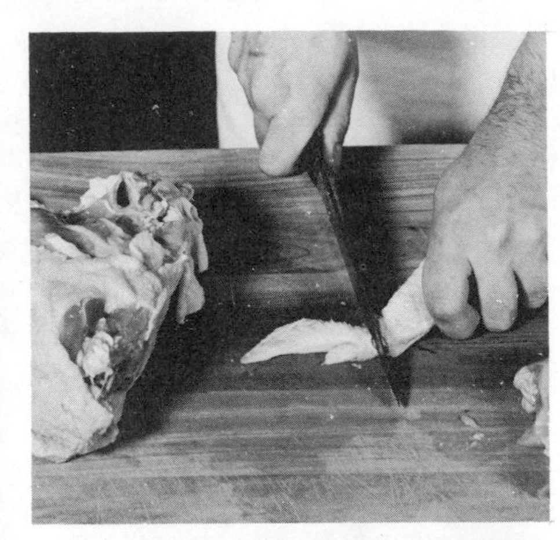

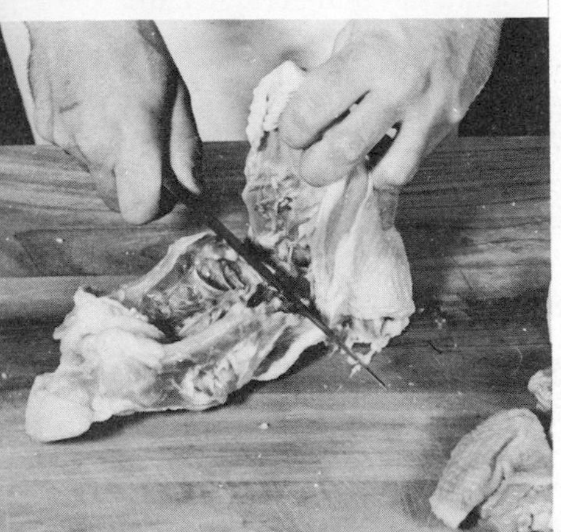

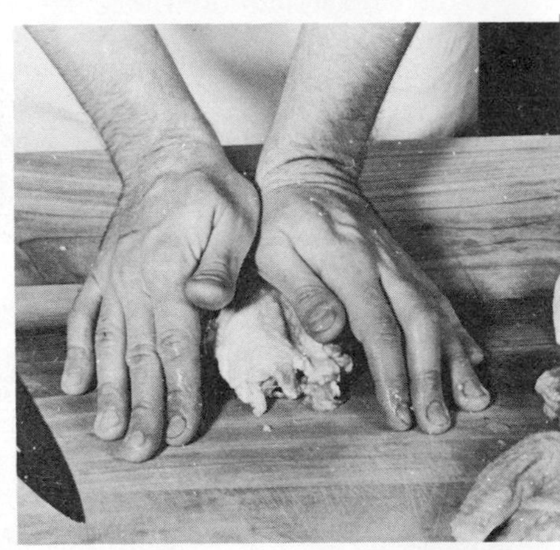

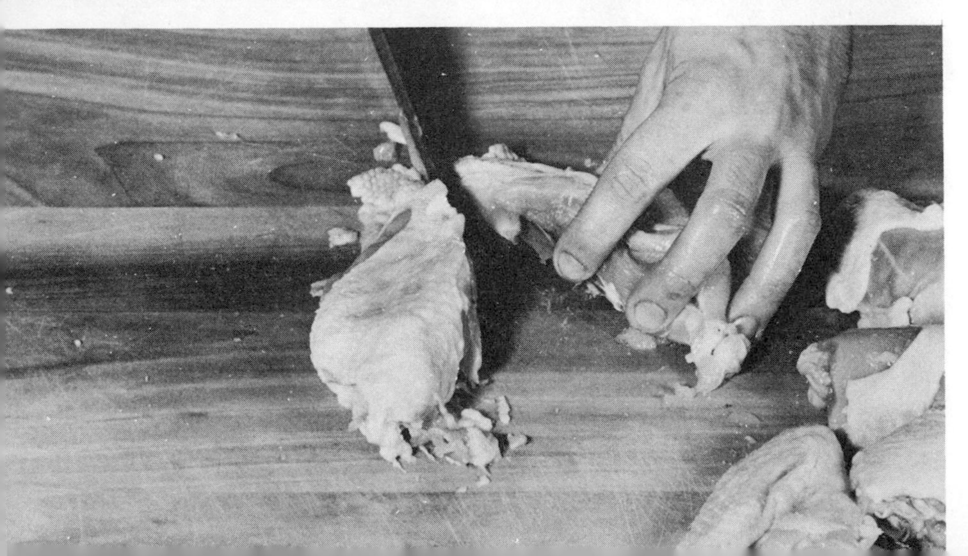

It is advisable to cut up poultry for freezing. You'll save much freezer space if you do, because you won't have the wasted space of the body cavity. You may bone or split the breast in half and leave the other pieces whole, or bone the whole bird for compactness. Roasters that are to be frozen whole should be trimmed of excess fat. Oxygen, which causes rancidity, is absorbed by fat, and by cutting off unnecessary fat you are retarding rancidity. To save space when freezing whole roasters, tie legs and wings tightly around the body of the bird. (For more information on freezing poultry, see the section on freezing meat.)

FREEZING MEAT

You may be able to do without a freezer for storing fruits and vegetables, but you'll more than likely find one indispensable when it comes to keeping meat, especially if you've got meat from a steer, calf, lamb, or hog to store for the good part of a year or more. Freezing is unquestionably the easiest and safest way to keep meat for long-time storage.

Most all meats freeze successfully, with the exceptions of processed and spiced meats, canned hams, and cured and smoked products. Luncheon meats, like salami, bologna, and spiced ham should not be frozen, but stored in the refrigerator and used within a week. Cured meats become rancid more quickly than meats that are frozen when fresh because the ingredients used in curing increase meat's ability to absorb oxygen. These products should not be kept frozen for more than one to two months (see time chart). The period homemade fresh sausage may be safely stored in the freezer can be lengthened if the spices are added after the meat is thawed. Seasonings and fillings for meat loaves, meat balls, and the like should be added after, not before, freezing and thawing ground meat for extended storage life. Seasonings limit freezer life.

If you are freezing, wrap your meat as soon as possible after it has been cut. Wrapping carefully is extremely important in preserving the flavor, texture, and freshness of all frozen food. Wrapping, weighing, and labeling meat from a large animal, like a steer, will require the work of at least two people if it is to be done in less than an eight-hour day.

WHAT WRAPS TO USE

There are a number of wraps on the market suitable to use for freezer storage. Heavy-duty plastic, aluminum foil, and freezer paper will protect frozen meats. Be certain that the wrap you use is moisture-proof. The air in freezers is relatively dry. If the wrapper is porous or poorly sealed, dry air will get in and draw moisture from the meat, dehydrating the surface and causing freezer burn. Although these burns are not harmful to meat, the dried area will have an unappetizing color and be tough and tasteless when cooked. This is particularly detrimental to foods with high moisture contents, like meats.

Your freezer wrap should be pliable so that it will mold itself to irregularly shaped meats and eliminate as much air as possible. Oxygen from the air which is absorbed by the meat will hasten the rate of rancidity.

The wrap must also be strong. Wraps not made for freezer storage are often too weak to use. These weaker wraps are more likely to tear and allow oxygen and dry air enter when meat is put in, taken out, or shuffled around inside the freezer. Take special care with cuts of meat that have sharp corners or protruding bones. These may be protected from tearing by placing a plastic bag, stockinette, or old nylon stocking over the freezer wrap.

Meats that are highly perishable, like pork and cured meats should be wrapped in a double thickness of wrap if they are to be kept longer than five months. Beef, veal, and lamb need only be wrapped in single thickness if they are to be used within one year's time. If they are to be kept longer, it is advisable to cover them in wrap of double thickness. It is a good idea to place wax paper or plastic wrap between hamburgers, sausage patties, chops, and steaks so that individual frozen pieces may be separated and cooked separately. Some people wrap their

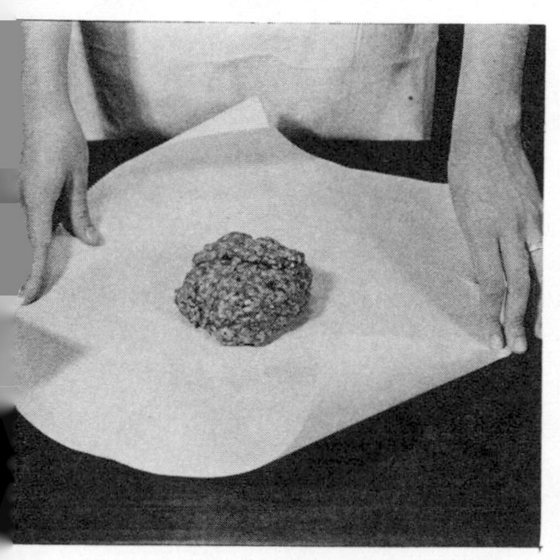

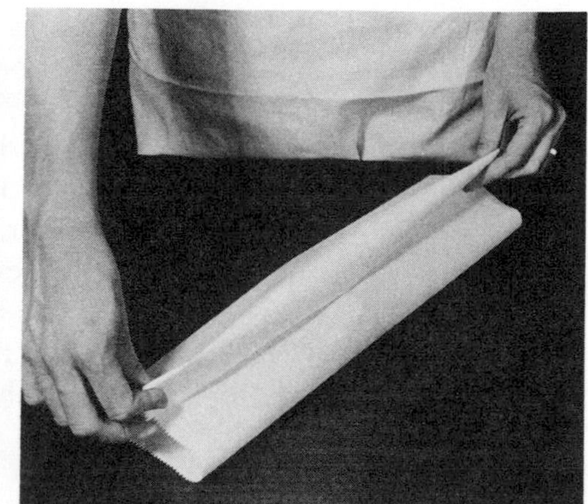

e best way to wrap meats for the freezer to make a delicatessen wrap. To make h a wrap, grab two ends of the freezer per, press them together, and fold them er several times. Then press the fold inst the meat to squeeze out all the air l make a compact package. Now fold up two loose ends of the paper, and seal paper securely with freezer tape. sking tape may seal your wrap, but it a tendency to lose its adhering ability e it is exposed to freezing tempera- es or moisture. Remember to label your kages.

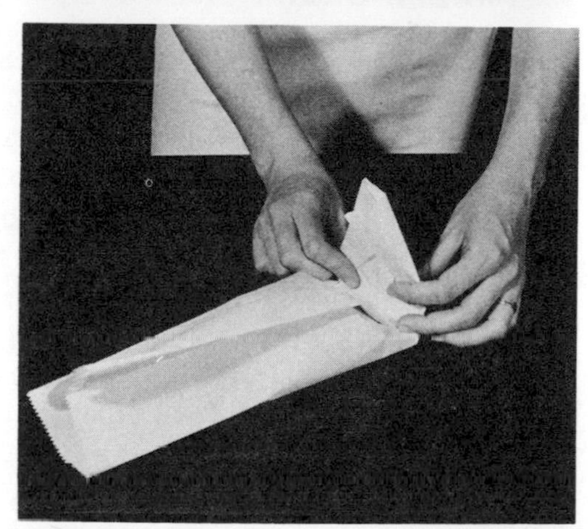

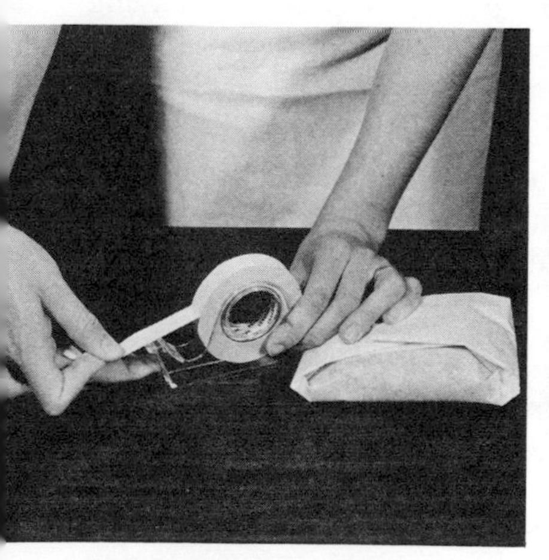

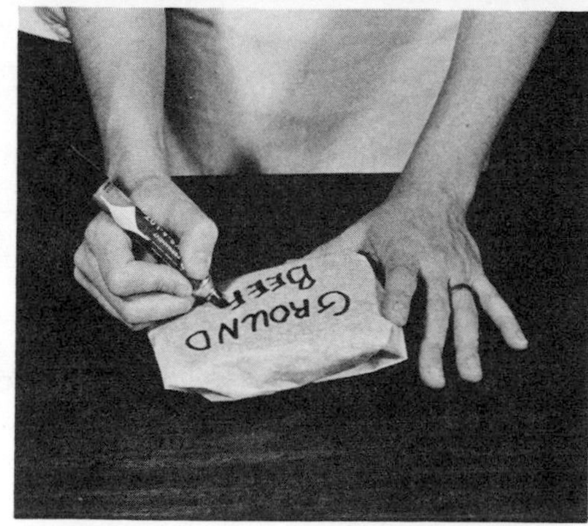

meats in plastic before they wrap them in freezer paper. This insures an airtight, waterproof seal. All bloody cuts, like the organ meats, should always be wrapped first in plastic and then in freezer paper.

After wrapping, every cut should be weighed (a kitchen scale will do nicely), and the weight, type of cut, and date frozen should be marked on the wrapper. With proper labeling, you'll be able to go to your freezer a few weeks or months later and choose the size and cut you want at a glance. As with any frozen food, use it in the order in which it was frozen. Cuts labeled with the earliest freezing dates should be used first, whenever practical.

FREEZING SOUP STOCK

The juices in meat bones may be extracted by simmering the bones in water to make soup stock. Bones should first be cracked or crushed to free juices and gelatinous matter. Marrow bones make for a better stock, but if too many are used the stock will be too gelatinous and have a thick, gluey consistency. Reserve such stock for sauces and gravies. Make a concentrated stock by adding just enough water to cover the bones. Add spices, if desired, and simmer the stock in a heavy pot with lid tilted at an angle so that the pot is practically covered. Simmer for at least twelve hours. Cool the stock and skim off the fat with a ladle. Fat remaining in the stock will hasten rancidity. Once skimmed, strain the stock through a double layer of cheesecloth and pour stock into plastic containers or heavy glass jars, leaving a half-inch headspace for expansion. Cover tightly and freeze. When you are ready to make soup from your frozen stock, thaw the stock and water it down to desired strength. If you are adding meat, simmer it in the stock until almost tender, and then add your vegetables. Simmer again just until meat is tender and vegetables are chewy, but neither hard nor soft.

QUICK-FREEZING

For best quality, meat should be frozen quickly. Slow freezing gives water within the meat tissues time to separate out and form large ice crystals which stretch and rupture surrounding tissues. Meat frozen rapidly results in little water separation and smaller ice crystals that do little damage to the meat tissues. Butchers have a special freezer that maintains a very low temperature to flash-freeze meat. Many home freezers have quick freeze compartments to freeze fresh meat solid. Once the meat is completely frozen in this quick freeze section, it is then transferred to the regular freezing compartment which will keep it frozen at about zero degrees. If your freezer doesn't have this special compartment, but it does have a temperature control, turn it to the coldest position and wait twenty-four hours after the fresh meat has been placed in the freezer before turning the control back to storage position. Don't overburden your freezer by putting a large amount of fresh meat into it at any one time. The amount of fresh meat placed in the freezer at one time should not exceed two to three pounds per cubic foot of freezer space in a twenty-four hour period. More than this amount raises the temperature and slows the freezing process. When adding fresh meat to your freezer, put it in the coldest parts; this is usually along the bottom and walls of the freezer unit. To aid your freezer in freezing meat quickly, pack fresh meat loosely so that the cold air can circulate freely around your cuts and freeze the meat rapidly. After the meat is frozen solid, you can repack it tightly to make the most of your freezer space.

Freezer compartments inside regular refrigerators are seldom cold enough for long-time storage of meats. Use only a chest or upright freezer or the separate freezer compartment of a two-door refrigerator-freezer. Or rent a freezer locker at a freezer plant. Lockers may be rented by the month, quarter, half, and full year at reasonable rates. Temperatures in such plants are kept at zero degrees or below, and some plants employ butchers who will cut and wrap meat to order. Although the number of freezer locker plants has shrunk in recent years due to the popularity of home freezers, there are still thousands in operation throughout the United States.

Should your freezer stop operating and you are unable to keep temperatures in it below freezing, a freezer locker can hold your meat until your freezer resumes operation. What to do in such emergencies is discussed in the chapter on freezing fruits and vegetables.

THAWING AND COOKING FROZEN MEAT

Meats need not be thawed before cooking. There is little or no difference between the quality of meat thawed before cooking

FREEZER STORAGE TIME CHART

Freezer Temperature (0°F. or Colder)

MEAT (in Freezer Wrapper)	Recommended Maximum Storage Time
Fresh meat	
Beef	6 to 12 months
Veal	6 to 9 months
Pork	3 to 6 months
Lamb	6 to 9 months
Ground beef, veal and lamb	3 to 4 months
Ground pork	1 to 3 months
Variety meats	3 to 4 months
Sausage and Ready-to-serve	
Luncheon meats	not recommended
Sausage, fresh pork	2 months
Frankfurters	1 month
Cured, Cured and smoked	
Bacon	1 month
Smoked ham, whole or slices	2 months
Beef, corned	2 weeks
Cooked Meat	
Leftover cooked meat	2 to 3 months
Frozen Combination Foods	
Meat pies	3 months
Swiss steak	3 months
Stews	3 to 4 months
Prepared dinners	2 to 6 months
Poultry	
Chicken (ready-to-cook)	6 to 7 months
Turkey (ready-to-cook)	6 to 7 months
Cooked chicken	2 to 3 months
Cooked turkey	2 to 3 months

Courtesy of Home Economists in Business

and meat cooked frozen. If you do cook meat when still frozen, you'll have to be more patient because frozen meat takes more time to cook than fresh or thawed. While thin steaks take about the same time to broil or panbroil, frozen or thawed, larger cuts like roasts take almost twice as long to cook.

When you do thaw meat, it is best to thaw it in its original freezer wrapper. In order to insure uniform defrosting throughout your piece of meat, defrost it at low temperatures—in the refrigerator if possible. You can also thaw meat in cold water, provided it is in a watertight wrapping. If you must thaw poultry at room temperature, take special precautions to keep the surface of the bird cool during thawing. Place the bird, still in its original wrapper, in a closed double bag until it is pliable. Don't let frozen meats, especially poultry or pork, thaw on surfaces or trays where other foods are kept.

TIMETABLE FOR THAWING FROZEN MEAT

Meat	Refrigerator	Low Temperature Oven (155°F.)
Large Roast	4 to 7 hrs. per lb.	1 to 1½ hrs. per lb.
Small Roast	3 to 5 hrs. per lb.	¾ to 1 hr. per lb.
1 inch steak	12 to 14 hrs.	1 hour

TIMETABLE FOR THAWING FROZEN POULTRY

Poultry	Ready-to-Cook Weight	In Refrigerator	Low Temperature Oven (155°F.)
Roasting	3 to 8 lbs.	1 to 2 days	2 to 3½ hrs.
	8 to 12 lbs.	1 to 2 days	3½ to 5 hrs.
	12 to 20 lbs.	2 to 3 days	5 to 7 hrs.
Frying Chicken, cut up	about 2 lbs.	4 hours	1¼ hrs.

Courtesy Home Economists in Business

Thawed meats that have not reached temperatures above refrigerator temperatures (35° – 40° F.) may be refrozen, although it is not generally recommended. Each time meat is refrozen, there is some deterioration of quality; the ice crystals tend to rupture the fibers, breaking down the texture and letting more juices escape. If meat is above refrigerator temperature do not refreeze. Thawed meat will keep in the refrigerator as long as fresh meat.

CANNING MEAT

While the popularity of canning meats in the home has diminished with the increased use of home freezers over the last thirty years, this method of food preservation, which was brought into the home nearly a century ago, is still used by many American families to preserve fresh and cured meats. Although more time, work, and equipment is needed to can, canning has some advantages over freezing. There is no freezer to maintain year round at zero degrees and no damage done to the food if there should be a power failure or freezer breakdown.

Because meat is much more vulnerable to destructive enzymes and toxic bacteria than are fruits and vegetables, great care must be taken during the preparation and process of canning meats. Before considering canning meat, please read the section on preparing meat for storage.

Use only good-quality meat. Fresh meat should be chilled to 40° F. after butchering. Lamb and beef should not be aged more than forty-eight hours. If meat cannot be processed right after the animal heat is gone, store it at temperatures of zero degrees or lower until canning time. Meat may be processed for canning while it is still frozen, but be sure to allow extra time for the meat to cook sufficiently. If you wish to thaw meat before processing, it is best done gradually at low temperatures until most of the ice crystals have disappeared. Poultry should be rinsed and drained before it is processed.

EQUIPMENT

Equipment needed for canning includes a cutting board or other smooth, clean surface, sharp knives, a kettle for boiling water, thermometer, tongs, pot holders, glass jars with lids and bands or tin cans, lids, and a sealer, and a pressure canner or pressure saucepan. All equipment should be scrupulously clean. Wash metal utensils in hot soapy water and rinse with boiling water. Scrub surfaces of wooden equipment with hot soapy water and a stiff brush, and rinse with boiling water.

For further protection against bacteria, wooden equipment should be disinfected with a home disinfectant. Prepare the disinfectant following the directions on container, and soak wooden boards and utensils in the solution for fifteen minutes. Wash disinfectant off well with boiling water. If you use this wooden equipment in daily food preparation, it should be scrubbed well and rinsed with boiling water before it is put away so that no particles of food remain to attract bacterial growth.

Like other low-acid foods—vegetables and dairy products—meat may contain toxic bacteria that cause botulism, a severe form of food poisoning. There is no danger of botulism if canned meats are processed at 240° F. for the required length of time. Only a pressure canner or pressure saucepan can reach this temperature practically in the home. No other method of cooking, using an open or covered pot or steamer without pressure, is safe to use with meat and foods containing meat, such as stews, soups, and gravies. Oven canning is impossible with tin cans and is hazardous with glass jars. A temperature of 240° F. inside the container cannot be reached. There is a chance that jars in the oven may burst, blowing out the oven door and causing considerable damage to the kitchen and to the person canning.

Before canning, make sure that your pressure canner is clean and is working properly. Remove the lid and wash the kettle in hot, soapy water. Do not wash the lid, but wipe it with a damp cloth. Clean the petcock and safety valve by running a string through the openings. Gauges should be checked. If you live in a high altitude area, increase the pressure by one pound for each 2,000 feet above sea level. If your canner has a dial gauge, it should be examined every year for inaccuracy by the manufacturer or dealer. Most all processing directions call for ten pounds pressure. The U.S. Department of Agriculture gives the following adjustments to correct the pressure on dial gauges:

If the gauge reads high –
1 pound high – process at 11 pounds.
2 pounds high – process at 12 pounds.
3 pounds high – process at 13 pounds.
4 pounds high – process at 14 pounds.

If the gauge reads low –
1 pound low – process at 9 pounds.
2 pounds low – process at 8 pounds.
3 pounds low – process at 7 pounds.
4 pounds low – process at 6 pounds.

Do not use a pressure canner with a gauge that registers as much as five pounds high or low.

If you are using glass jars, be sure that all jars, lids, and bands are in perfect condition. Discard any with cracks, nicks, or chips. Even slight imperfections may prevent proper sealing. Use new rubber rings each time you can. Jars need not be sterilized, but they should be washed in hot, soapy water and rinsed well with boiling water. Pint and quart jars are good for canning meats, but half-gallon jars take too long to process and are not recommended.

If you're using tin cans instead of jars, use plain tin cans, without enamel lining. Use only perfect, rust-free cans and lids. Cans should be washed, rinsed well, and drained just before packing. Don't wash the lids, but wipe them with a damp cloth if they are dirty. Washing may damage the gaskets. Make sure that your sealer makes an airtight, smooth, finished seam. It is a good idea to test the sealer before canning by sealing a can of water. Submerge the sealed can of water in boiling water for a few seconds and look for air bubbles. If they rise from the can, your sealer needs adjusting.

CANNING PROCEDURE

1. Prepare meat.

Beef: Trim off fat and de–bone to save space and make packing easier. Cut tender meats, like roasts, steaks, and chops, into pieces the length of the can or jar with the grain of the meat running lengthwise. Tougher pieces of meat should be ground or cut into chunks for stewing meat. Bony pieces may be used to make broth or stock for canning.

Poultry: Cut into container-sized pieces. Remove the bones from the meaty parts, like the breast. Separate the thighs from the drumsticks. Keep the giblets to process separately.

2. Make broth, if desired, to cover meat.

Place meat bones in lidded pot with water and simmer until meat is tender. Skim off fat and save broth for packing .

3. Pack meat.

Meats should always be packed loosely. Containers may overflow if contents are packed too tightly or too full. Hearts may be packed after precooking (hot pack) or they may be packed before cooking (raw pack).

The chart that follows, which was adapted from the U.S. Department of Agriculture Bulletin, "Home Canning of Meat and Poultry," will give you an idea of the number of glass jars you will need to can your cuts of meat.

YIELD OF CANNED MEAT FROM FRESH

Cut of Meat	Pounds of Meat per Jar	
	Pints	Quarts
Beef:		
Round	1½ to 1¾	3 to 3½
Rump	2½ to 2¾	5 to 5½
Pork Loin	2½ to 2¾	5 to 5½
Chicken:		
Canned with bone	1¾ to 2	3½ to 4¼
Canned without bone	2¾ to 3	5½ to 6¼

Hot Pack: Meat is cooked to at least 170° F. in a skillet, pot, or in the oven. It is then packed in cans or glass jars and boiling water or broth is poured over the meat, leaving a one-inch headspace in the container. Salt may be added to taste, if desired, but it is not necessary. Wipe the rim of the jar carefully and adjust closure.

If you are using a mason jar with two-piece metal cap, put the lid on so that the sealing compound is next to the glass rim. Screw metal band on tightly by hand. This lid now has enough give to let air escape during processing.

1. To can chicken in a hot pack, cut your bird into jar-sized pieces and cook them in broth or water until only a slight pink color remains. 2. Then pack the hot meat into clean jars by placing the thighs and drumsticks on the outside and the boned pieces in the center. Cover the poultry with boiling liquid, leaving a 1-inch headspace. 3. Screw the lids on tightly and set the jars on a rack in the pressure canner, allowing room around each one for steam to circulate freely. Fasten the lid on the canner and process for required time.

If you are using a mason jar with zinc porcelain-lined cap, fit the wet ring on the jar shoulder, but don't stretch it more than is necessary. Screw the cap down tightly and then turn it back one-quarter inch before processing.

If you are using cans, place the lid on each can with the gasket side down. Seal cans with sealer immediately, following manufacturer's instructions.

Raw Pack: Fill jars or cans with raw meat, leaving a one-inch headspace. Set open, filled containers on a rack in a pan of boiling water. Keep the water level two-inches from the top of the containers. Heat the meat slowly to 170° F., or for seventy-five minutes, if a meat thermometer is not available. Remove jars or cans from the pan and add salt to the meat, if desired. Wipe the rim of the container clean and adjust closure, according to directions above, under hot pack.

4. Processing.

a. Put two or three inches of water in the canner and heat water to boiling.

b. Set filled jars or cans on rack in the canner. Pack them into the canner loosely, allowing room around each one for steam to circulate freely. If there is room for two layers of jars or cans, place a rack between the two levels and stagger the containers so that none are directly over any of those below.

c. Fasten canner cover securely so that all escaping steam exits only through the petcock or weighted gauge opening.

d. Let pressure reach ten pounds (which is 240° F.). Note the time as soon as the gauge reads ten pounds and start counting processing time. If you're using a pressure saucepan, add twenty minutes to required processing time. Maintain ten pounds on the gauge by adjusting heat if necessary. Varying pressure may cause containers to overflow and lose some of their liquid.

e. When processing time is up, remove the canner from the heat immediately.

f. If you are using jars, do not pour cold water over the canner to reduce pressure quickly, but let the canner stand until the gauge reads 0° F. Wait a few minutes after gauge reads 0° F. and then slowly open petcock or take off weighted gauge. Unfasten the cover and tilt it away from you so that steam can escape without rising in your face. When all the steam is gone, remove the jars.

If you are using cans, remove the canner from the heat as soon as gauge reads 0° F. and open petcock or remove weighted gauge to release steam. Unfasten cover and tilt it away from you so that steam can escape without rising in your face. Remove cans.

5. Cool containers.

Glass Jars: As soon as they are taken from the canner, complete seals, if necessary. Only mason jars with zinc porcelain-lined caps need to be sealed. Do this by quickly screwing the cap down tightly. Mason jars with two-piece metal caps are self-sealing. Cool the jars right side up on a rack or folded cloth. Don't cover them, and don't cool them in a drafty place.

Tin Cans: Put them in cold water as soon as they are removed from the canner. Change the water frequently for rapid cooling. Remove the cans from water while they are still warm and allow them to dry in the air. If the cans are stacked, stagger them so that the air can circulate freely around them.

6. Check your seals.

When containers are thoroughly cool, examine each carefully for leaks. Press on the center of the jar lid. If this lid does not "give" when you press on it, the jar is sealed. Check your cans by examining all seams and seals. Can ends should not be bulging, but almost flat; and seams should be smooth with no buckling. If you suspect a can or jar of having a faulty seal, open the container, reheat the meat, and process all over again, for the complete time required, to insure safety. Jars that have lost liquid during canning should not be opened and reprocessed. Although the meat inside such cans may darken during storage, the meat is not spoiled.

7. Mark jars.

Write directly on the top of the jar lids, or use adhesive tape, freezer tape, or special labels for cans and jars. Label each container with the type of food canned and date of canning. Canned foods should be used in the order they were canned.

8. Store the containers.

Jars and cans that contain meat should be stored in a cool, dry place. Do not subject the canned foods to warm temperatures or direct sunlight, as they will lose quality. Freezing does not cause canned meat to spoil, but it may damage the seal so that spoilage begins. To protect against freezing in an unheated area like a cellar or garage, cover the jars and cans with a clean blanket or wrap them separately in newspapers. A damp storage area invites rust which corrodes cans and metal jar lids and causes leakage.

Meat that has been processed, sealed, and stored properly will not spoil. If you suspect meat of being spoiled, don't test it by tasting. Destroy it by burning or dispose of it where it cannot be eaten by animals or humans. It is a good idea to boil all home-canned meat twenty minutes in a covered pot before tasting or serving it. Twenty minutes of rapid boiling will destroy any dangerous toxins that remain in foods that were improperly processed. This precaution should certainly be followed by anyone who is canning for the first time or is not certain that his gauge is accurate. Boiling is the best way to find out if meat is safe to eat because heat intensifies the characteristic odor of spoiled meat. If your meat develops an objectionable odor, dispose of it without tasting.

Jars and cans should be checked during storage and before opening for signs of spoilage. Any abnormality in your can or jar—bulging jar lids or rings, gas bubbles, leaks, bulging can ends—can mean spoilage. If you notice any of these signs, dispose of the container and contents without tasting. If you notice an off-odor, discoloration, or spurting liquid as you open a can or jar you know that the food is not safe to eat. If there is a discoloration on the metal lids or cans, this is most likely caused by sulfur in the meat; this does not mean the meat is spoiled.

Directions for Processing Meat

Type and cut	Preparation and processing	Processing Time
Cut-up beef, veal, lamb or pork	Use tender cuts for canning in strips. Making sure that grain of meat runs lengthwise, cut meat the length of the can or jar. Cut less tender cuts into cubes for stewing or soups.	
	Glass Jars: Hot Pack – Precook meat in skillet or saucepan in just enough water to keep meat from scorching. Stir occasionally so that all pieces heat	

evenly. Cook until medium done. Pack hot meat loosely, leaving a 1-inch headspace. Add ½ teaspoon salt to pint jars and 1 teaspoon to quarts, if desired. Cover the meat with boiling water or boiling juice from meat, leaving a 1-inch headspace. Adjust lids and process in pressure canner at 10 pounds pressure for required time.

Quarts
90 min

Pints
75 min

Raw Pack – Pack cold, raw meat loosely, leaving a 1-inch headspace in jar. Exhaust air by heating meat-filled jars in a pot of hot water. Cook at a slow boil until meat is medium done, or reaches 170° F. Add ½ teaspoon salt to pints and 1 teaspoon salt to quarts, if desired. Adjust lids and process in pressure canner at 10 pounds for required time.

Quarts
90 min

Pints
75 min

Tin Cans:
Hot Pack – Precook meat in skillet or saucepan in a small amount of water until meat is medium done, stirring occasionally so that meat heats evenly. Pack cooked meat in cans, leaving a ½-inch headspace. Add ½ teaspoon of salt to No. 2 cans or ¾ teaspoon salt to No. 2½ cans, if desired. Fill cans to top with boiling water or boiling juice from meat. Seal. Process in a pressure canner at 10 pounds for required time.

No. 2
65 min

No. 2½
90 min

Raw Pack – Pack cans with cold, raw meat to top. To exhaust air, cook meat in cans in a pot of hot water. Cook at a slow boil until meat is medium done or reaches 170° F., about 50 minutes. Add ½ teaspoon salt to No. 2 cans and ¾ teaspoon salt to No. 2½ cans, if desired. Press meat down into can so that there is a ½-inch headspace and add boiling water or broth to fill to top. Seal cans. Process in a pressure canner at 10 pounds pressure for required time.

No. 2 65 min

No. 2½ 90 min

Ground meat

Glass Jars:

Hot Pack – If you are making patties, press the meat tightly into patties that will fit into your jars without breaking. Meat may also be left loose although it is harder to remove from jars. Precook patties in a slow oven or over low heat in a skillet until medium done. Pour off all fat; do not use any fat in canning. Pack patties or loose meat in jars, leaving a 1-inch headspace. Cover with boiling water or broth to cover meat, leaving 1-inch headspace. Adjust lids and process in pressure canner at 10 pounds for required time.

Quarts 90 min

Pints 75 min

Raw Pack – Pack raw ground meat into jars, leaving 1-inch headspace. Cook meat-filled jars in a pot of hot water at a slow boil until medium done or to 170° F., about

Quarts 90 min

75 minutes. Adjust jar lids and process in a pressure canner at 10 pounds pressure for required time.

Pints
75 min

Tin Cans:

Hot Pack – If you are making patties, press the meat tightly into patties that will fit into your cans without breaking. Meat may also be packed loose. Pack loose meat or patties into within ½-inch of top of can. Precook patties in slow oven or over low heat in a skillet until medium done. Pour off all fat. Cover with boiling water or boiling juice from meat to the top of the cans. Seal and process in a pressure canner at 10 pounds pressure for required time.

No. 2
65 min

No. 2½
90 min

Raw Pack – Pack raw ground meat solidly to the top of the can. Cook meat in pot of hot water at a slow boil until medium done or 170° F., about 75 minutes. Press meat down into cans to allow for a ½-inch headspace and seal. Process in pressure canner at 10 pounds pressure for required time.

No. 2
100 min

No. 2½
135 min

Sausage Sausage may be packed and processed just as ground meat. Follow sausage recipes in "Preparing Beef, Veal, Lamb and Pork For Storage." Use seasonings sparingly in sausage intended for canning, as spices change flavor in storage. If you normally use sage, omit it; it makes canned sausage bitter.

Poultry, small game and rabbits

Sort meat into meaty and bony parts and can separately.

Glass Jars:

Hot Pack, with bone – Bone the breast and cut bony parts into jar-sized pieces. Trim off excess fat. Heat pieces in pan with broth or water. Stir occasionally until meat is medium done, when only slight pink color remains. Pack meat loosely, with thighs and drumsticks on the outside of the jar and boned pieces in the center. Leave a 1-inch headspace. Add ½ teaspoon salt per pint and 1 teaspoon per quart, if desired. Cover poultry with boiling water or broth, leaving a 1-inch headspace. Adjust lids and process in pressure canner at 10 pounds pressure for required time.

Quarts 65 min

Pints 75 min

Hot Pack, without bone – Remove meat, but leave skin on meaty parts. Cook meat as above. Pack jars loosely, leaving a 1-inch headspace. Add ½ teaspoon salt per pint and 1 teaspoon per quart, if desired. Pour in boiling broth or water, leaving a 1-inch headspace. Adjust lids and process in pressure canner at 10 pounds pressure for required time.

Quarts 90 min

Pints 75 min

Raw Pack, with bone – Bone the breast and cut other pieces, with bone in, into jar-size pieces. Trim off excess fat. Pack raw poultry loosely by placing thigh and drumsticks on the outside and breasts in the center of the jar. Leave a 1-inch headspace. Place filled jars in a pan of hot water and cook at a slow boil until medium done, about 75 minutes, or until meat is medium done. Add salt if desired: ½ teaspoon per pint and 1 teaspoon per quart. Adjust lids and process in pressure canner at 10 pounds pressure for required time.

Quarts
75 min

Pints
76 min

Raw Pack, without bone — Remove bones, but not skin, from meaty pieces. Pack loosely, leaving a 1-inch headspace. Place filled jars in pan of hot water and cook at a slow boil until meat is medium done, about 75 minutes, or until thermometer reads 170° F. Add salt, if desired: ½ teaspoon per pint and 1 teaspoon per quart. Adjust lids, and process in pressure canner at 10 pounds pressure for required time.

Quarts
90 min

Pints
75 min

Tin Cans:

Hot Pack, with bone — Bone the breast and cut bony pieces into can-size pieces. Trim off excess fat. Heat pieces in a pan with broth or water. Stir occasionally until meat is medium done, when only slight pink color remains. Pack cans, leaving a ½-inch headspace. Add ½ teaspoon salt to No. 2 cans and ¾ teaspoon salt to No. 2½ cans, if desired. Fill cans to the top with boiling broth. Seal and process in

No. 2
55 min

a pressure canner at 10 pounds pressure for required time.

No. 2½ 75 min

Hot Pack, without bone – Remove bones, but not skins, from meaty pieces. Cook as above. Pack loosely, leaving a ½-inch headspace. Add salt, if desired: ½ teaspoon per No. 2 can and ¾ teaspoon per No. 2½ can. Fill cans to the top with boiling broth and seal. Process in a pressure canner at 10 pounds pressure for required time.

No. 2 65 min

No. 2½ 90 min

Raw Pack, with bone – Bone breast and cut thighs and drumsticks into can-sized pieces. Trim off excess fat. Pack poultry loosely, with thighs and drumsticks on the outside and breasts in the center. Pack to the top of the can. Place filled cans in a pan of hot water and boil slowly until meat is medium done, or until thermometer reads 170° F. Add salt if desired: ½ teaspoon per No. 2 can and ¾ teaspoon per No. 2½ can. Seal and process in a pressure canner at 10 pounds for required time.

No. 2 65 min

No. 2½ 90 min

Raw Pack, without bone – Cut poultry and remove bone, but not skin. Pack raw meat to top of the cans. Place filled cans in a pan of hot water and slowly boil until meat is medium done, about 75 minutes, or until thermometer reads 170° F. Add salt, if desired: ½ teaspoon per No. 2 can and ¾ teaspoon per No. 2½ can. Seal and process in a pres-

No. 2 65 min

	sure canner at 10 pounds pressure for required time.	No. 2½ 90 min
Giblets	Can in pint jars or No. 2 cans. Separate hearts and gizzards from livers and cook and can them separately to avoid blending of flavors.	
	Hot Pack – Cook giblets in a pan with water or broth until medium done. Stir occasionally so that meat heats evenly.	
	Glass Jars – Pack meat, leaving a 1-inch headspace. Add boiling water or broth just to cover giblets. Adjust lid and process in a pressure canner at 10 pounds pressure for required time.	Pints 75 min
	Tin Cans – Pack giblets, leaving a ½-inch headspace. Fill cans to top with boiling broth or water. Seal and process in a pressure canner at 10 pounds for required time.	No. 2 65 min
Corned beef	Use the recipe given in "Storing Beef, Veal, Lamb and Pork for Storage." Wash and drain corned beef. Cut it into container-sized strips. Cover the meat with cold water and bring to a boil. Taste broth; if it is very salty, drain and boil meat in fresh water. Pack while hot.	
	Glass Jars – Pack meat loosely, leaving a 1-inch headspace. Cover meat with boiling water or broth, leaving a 1-inch headspace. Adjust lids and process in a pressure canner at 10 pounds for required time.	Quarts 90 min Pints 75 min

<table>
<tr><td></td><td>Tin Cans – Pack meat loosely, leaving a ½-inch headspace. Fill cans to top with boiling water or broth. Seal and process in a pressure canner at 10 pounds pressure for required time.</td><td>No. 2
65 min

No. 2½
90 min</td></tr>
<tr><td>Liver</td><td>Wash, remove skin and membranes. Slice into container-size pieces. Organ meats are hot packed. Drop into boiling water for 5 minutes. Follow hot pack directions for cut-up meat.</td><td></td></tr>
<tr><td>Heart</td><td>Wash and remove thick connective tissue. Cut into container-sized strips or 1-inch cubes. Put in a pan and cover with boiling water. Cook at a slow boil until medium done. Follow hot pack directions for cut-up meat.</td><td></td></tr>
<tr><td>Tongue</td><td>Wash, put in pan and cover with boiling water. Cook for about 45 minutes, until skin can be removed easily. Skin and slice into container-size pieces or into 1-inch cubes. Reheat to simmering in the same boiling water. Follow hot pack directions for cut-up meat.</td><td></td></tr>
<tr><td>Soup stock or broth</td><td>Crack or saw bones. Simmer them in water until meat is tender. To save work and storage space, make a concentrated stock which can be diluted before reheating. Simmer bones in just enough water to cover them. Strain stock and skim off fat. Cut up meat and return it to stock, if desired. Reheat to boiling and</td><td>Quarts
25 min</td></tr>
</table>

pack into glass jars. Adjust lids and process in a pressure canner at 10 pounds pressure for required time.	Pints 20 min

CURING AND SMOKING MEATS

Our chief concern in this book is to explain how best to preserve food for optimal food value. This means preserving as much of the original supply of nutrients in food as possible by giving great attention to the many steps involved in harvesting, processing, and storing foods. It also means being careful not to add anything to our stored foods that is harmful to our bodies, like adding For this last reason we must begin this chapter on curing and smoking meats with a few words of caution.

There are a number of things that we find objectionable in cured and smoked products. Basically, curing meat means salting it so much that it will resist the bacteria that would otherwise cause the meat to spoil. Anyone who has ever tasted bacon, dried beef, or country ham knows that a lot of salt has been absorbed by the product. And salt, especially in the great quantities that you'll find in cured products, doesn't do our bodies any good. Too much salt can be the cause of hyperacidity, high blood pressure, and excessive water retention. It may interfere with the absorption and utilization of food by over-stimulating the digestive tract, and it can aggravate heart disease.

Salt is the basic curing ingredient, but if used by itself it tends to make meat oversalty, dry, and tough. So along with salt, most cures contain sugar. Our objections to sugar, discussed in earlier sections of this book, hold true for cured and smoked meats also.

Beef and pork cures may contain saltpeter as a color fixative. We object to the consumption of saltpeter because it has been proven to be toxic to man. The nitrate in saltpeter combines easily with amines in food, drugs, alcohol, and tobacco smoke to produce nitrosamines, and these compounds have been found to cause cancer.*

Smoked meat tastes smokey because the meat has actually absorbed smoke emitted from slow-burning hardwoods. The smoke is a coal-tar derivative, and coal-tar derivatives have long been known as cancer-causing substances.

We recommend freezing or canning meat before we advise curing and smoking it. But, if your homestead doesn't have the facilities for freezing or canning meat, you'll have to rely on curing, and maybe smoking it too, if you want to keep it for any length of time. The salt in the cure extracts water and checks the growth of most spoilage organisms. The smoke colors and flavors the meat, dehydrates the cured meat further, and slows down the development of rancidity in the fat. However, meat that has been cured properly does not also need to be smoked to insure good preservation.

There aren't too many old-time butchers and smokers left; they've gone out with the introduction of commercially prepared cures and automatic smokehouses. Luckily, here in Pennsylvania Dutch country you can still find a few country butchers who are doing their own smoking—just as it's been done for one hundred and more years. Thomas Meyers, of Fleetwood, Pennsylvania, is one of these men. Some sixty years ago his father taught him the art of smoking, and he's been smoking meats the same way ever since. His smoked bacon, ham, beef, and chicken have an aroma and taste distinctively different from the smoked foods sold by the bigger, modern operations that forced him to give up his business a few years ago. Now he has a small shop and smokehouse which serves customers of a popular country restaurant.

Mr. Myers is a friendly man and was glad to talk to us about

*I.A. Wolff, "Nitrosamines," *Journal of Environmental Health* 35:2 (September/October, 1972): pp. 114–18; and S. Govindarajan, "Nitrities and Nitrates in Food, "*Food Product Development* (October, 1972): pp. 33–4.

his trade, but when we asked him just how he smokes his products he had a difficult time explaining the procedures to us. "I can't tell you; I just do it," was his reply. "You'd have to see for yourself." He didn't learn how to smoke from a book; he learned by working with his father and practicing on his own until he developed a real eye and feel for his work.

And this is just what you'll have to do if you want to produce a good smoked product. We can give you fine old, traditional recipes for cures, tell you how to build smokehouses and provide you with a set of general instructions, but the thing that we can't give you here in this book is the practice and know-how that can make the difference between a fair-tasting ham or side of bacon and a great-tasting one. This will have to come from you. If you're serious about curing and smoking your own meats, find yourself a small country butcher—one who takes great pride in his own meats—and work with him, at least once, when he's curing and smoking.

Before you decide if you are going to cure and smoke your own meat, ask yourself if you can build, buy, or borrow a smokehouse. It is the most important—and most expensive—piece of equipment you will need.

BUILDING THE SMOKEHOUSE

The first thing that we're going to discuss is the smokehouse, because a smokehouse is the most important and most complicated piece of equipment that you'll need if you intend to preserve meat by both curing and smoking. It's fairly easy to build a smokehouse out of just about anything—an old refrigerator, a packing crate, a barrel—providing it has four essential things: a source of smoke, an area in which to confine the smoke, racks to hold the meat, and a draft.

The smoke is usually produced by burning hardwood, a charcoal fire with hardwood chips or sawdust sprinkled over it, or even by burning corncobs. Never use softwoods; their pitch gives an off-odor and taste to the food.

A draft is absolutely essential to the smoking operation; with-

In a cement-block smokehouse, movable bricks near the roof may be used to vary the draft.

out it the smoke will stagnate in the smokehouse, and the smoked meat will have an objectionable, sooty taste. In a barrel smokehouse, the draft can be created by a plywood cover with a hole cut in it. In a more permanent arrangement, such as a cement-block smokehouse, movable bricks can be used to vary the draft. If it is possible, construct your smokehouse so that there are draft holes in the bottom of the smokehouse as well as the top. With draft holes at both ends you'll be able to control the circulation of the smoke more efficiently.

A Barrel Smokehouse

A fifty-gallon barrel can be made into a temporary smokehouse to accommodate a small amount of meat:

Dig a pit smaller than the diameter of the barrel and about two feet deep. Place the barrel, with the top removed and a hole cut in the bottom, over it. A fire pit should be dug ten to twelve feet away from the barrel in the direction of the prevailing winds. The pit should be deep enough so that it can be covered. The intensity of the fire in the pit and the amount of draft can be controlled by placing a piece of sheet metal over the pit. The pit should be connected to the bottom of the barrel with a trench made from a stovepipe or sewer tile. It is important that this

You can build a simple type of smokehouse from a barrel. Build your barrel smokehouse outside on open ground, away from children, pets, and livestock.

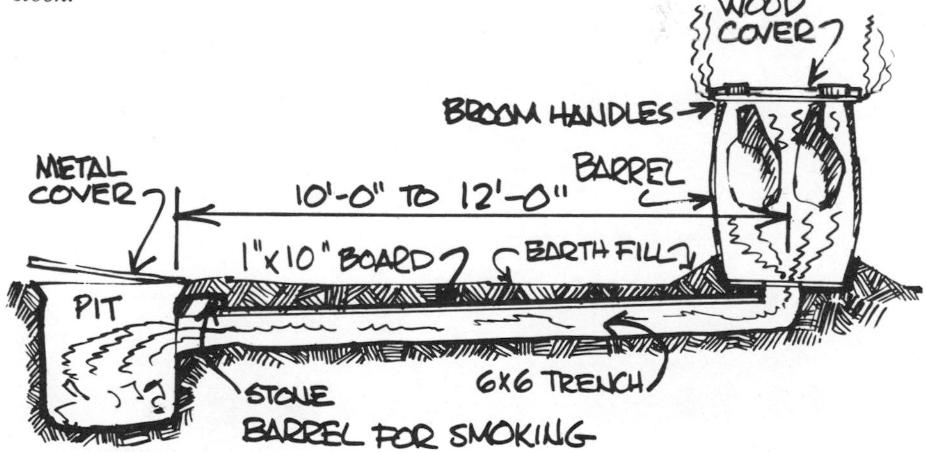

trench be airtight so that little smoke will be lost. A lid for the barrel should be made from plywood. Strips of wood or broom handles to support the hanging meat should be placed across the top of the barrel, as shown in the diagram below. The wooden cover is then placed over the wood strips. The openings made by the wood strips provide draft holes for escaping smoke.

Converting a Refrigerator into a Smokehouse

An old refrigerator can also be converted into a smokehouse. Because it is more durable than a barrel and is insulated, it will provide a more permanent smokehouse that retains heat well. In addition, its removable metal racks and full door make it very convenient for hanging and removing meat.

An old refrigerator can be converted into a more permanent smokehouse. Be sure to place the pit at least 10 feet from the refrigerator to prevent insulation in the refrigerator walls from catching on fire.

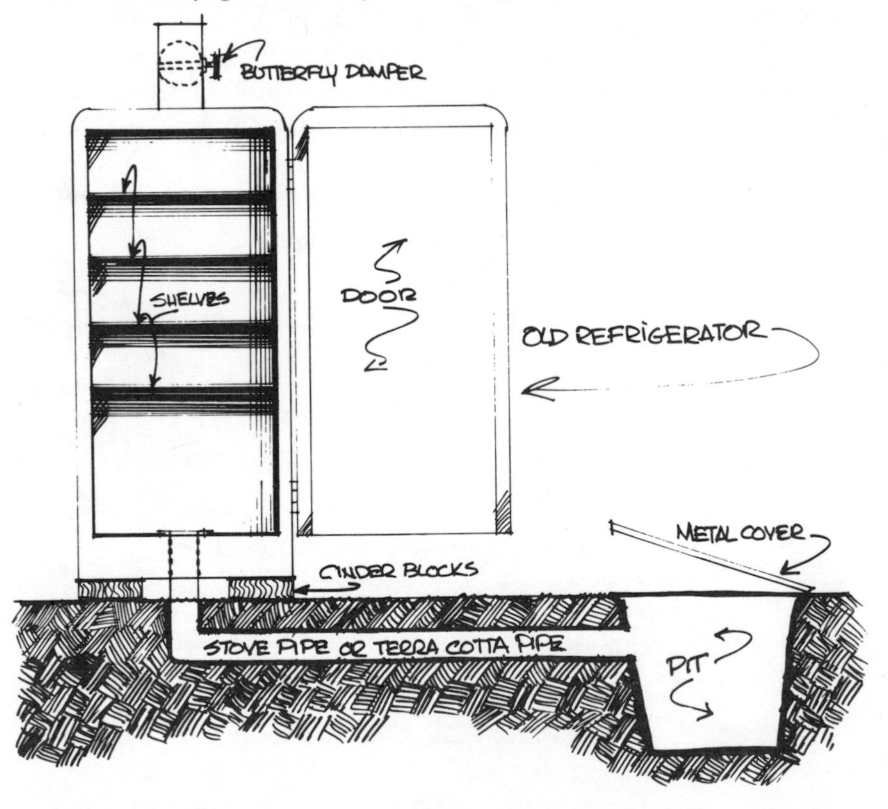

Although the heat can be placed directly underneath the refrigerator, or even inside it, this arrangement is not advisable because the insulating material inside the refrigerator walls may catch on fire if it is subjected to direct heat. Rather, it is best to dig a fire pit several feet from the refrigerator unit and direct the heat to it by using an underground trench, as we recommended for the barrel smokehouse.

If the motor and compressor are contained in the bottom of the refrigerator, remove these. Place the unit on four cinder blocks, cut a hole in the bottom and attach a tight-fitting stovepipe or sewer tile leading to the fire pit from this hole to prevent smoke from escaping. Provide a draft at the top of the unit by cutting a three-inch hole in the top and attaching a two-foot length of stovepipe with a butterfly damper.

This type of conversion will allow you to make maximum use of the refrigerator space and prevent any chance of the insulation from catching on fire.

A PERMANENT SMOKEHOUSE

A permanent smokehouse can be built out of wood, cement blocks, or concrete, and can be modified to suit your taste. One such smokehouse is shown below. It is sufficient for smoking meats of eight to twelve hogs. Pipes or wooden rods are hung on the wooden supports, and from these pipes the meat hangs. The oven behind the smokehouse can be used for rendering lard, or for cooking scrapple, if you're doing your own butchering. When the set-up is to be used for smoking, water is placed in the kettle, damper B is closed and damper A is opened. When the kettle is to be used for cooking, A is shut and B is opened.

It is also possible to modify an exisiting shed by adding a firebox at one end of the building. If a frame shed is used, it is advisable to construct a firebox outside the shed for safety. Dig an underground trench leading from the firebox to the center of the shed floor for the smoke. The trench can be made from stovepipe or sewer tile.

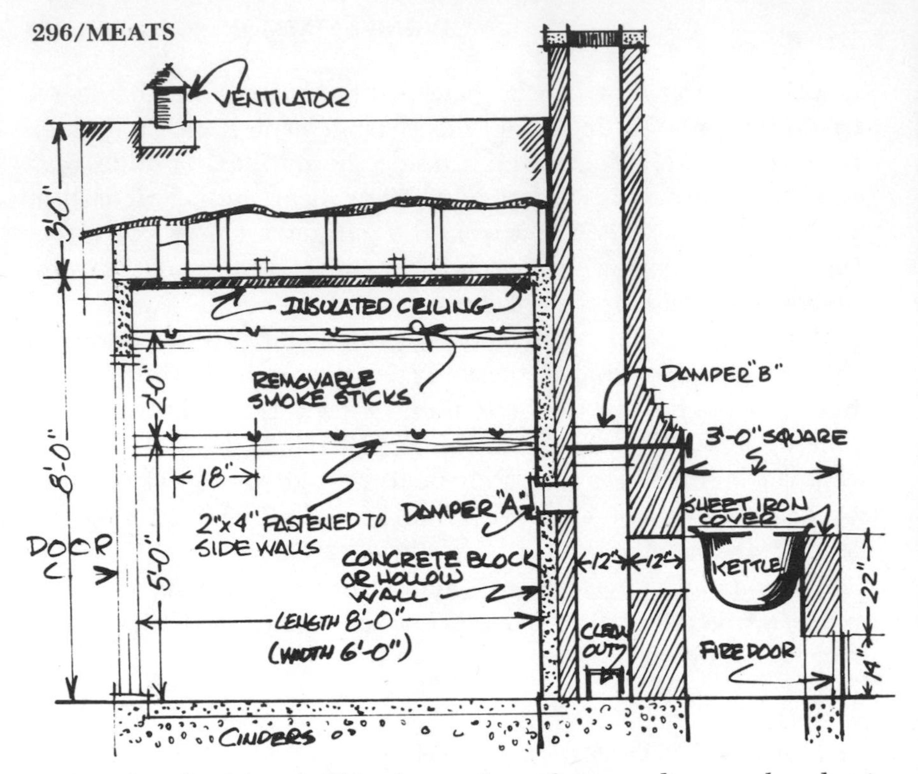

This 6-by-6-by-8-foot combination smoke and storage house and cooker is large enough to smoke the meat from 8 to 12 hogs at one time.

When to Cure and Smoke

It is vital to hold the temperature of the meat down while it is being cured: the temperature should not go over 38 – 40° F. Temperatures greater than this will cause the meat to spoil, and lower temperatures retard the salt penetration. This is where a refrigerator or spring house comes in handy, especially in warm weather. Most farmers, if given the choice, do their butchering in the fall and curing and smoking in the early winter. Cool weather helps to keep meat chilled when it is most important. By the time warm weather rolls around again, the meat is well aged.

Equipment for Curing

For curing you'll need a sturdy container for holding the meat. Glass or earthenware containers are excellent. Many farmers use large crocks. Never use aluminum containers. Wood containers aren't acceptable unless you are using a dry cure. A hardwood barrel will be all right, so long as it is watertight. If the barrel is new or has not been filled with a liquid for a long time, fill it with water and allow the water to set in it for at least twenty-four hours before you pour it out and use it for brining. The water will cause the wooden boards to swell and make the barrel watertight. If you are brine-curing, the container should be large enough to hold the meat and brine.

CURING

There are two types of cures: the dry cure and the brine cure. If you wish to make a brine cure, be certain that your water is clean and has a pleasant taste. If it is heavily chlorinated, we suggest that you boil the chlorine out of it first. Don't use bleach or any other chemicals to purify water; use bottled water instead. If you are unable to get good tasting, clean water, you'd do better to use the dry cure. Dry curing is usually the more successful of the two methods for the beginner and the person who does not have a refrigerated curing room. Since the dry cure is applied directly to the meat in its full strength, the rate of cure is more rapid than the brine cure. This is especially important in the southern parts of our country where the weather is warm most of the year. With this cure there is less chance of spoilage if the temperature should vary somewhat during the curing period. Curing time is not as exacting as it is with the brine cure, and there is no need for a brine pump, or watertight barrel or crock in which to soak the meat.

Commercially prepared cures are available from some salt companies and chemical manufacturers. Although these ready-made mixes are easy to use because they are pre-measured and premixed, they generally contain antioxidants. Most companies offer a choice of cures, but you have much more control over taste and degree of saltiness when you mix your own because you can regulate the amount of each ingredient.

Many recipes, for cures, even old-time recipes, use both sugar and saltpeter. Some use molasses instead of sugar, and some require both sugar and molasses. We haven't found anyone yet who uses honey, but the butchers that we have questioned see no reason why honey couldn't be used instead of sugar and molasses. The only drawback may be its high price (unless you keep your own bees). Most people agree that one-half the amount of honey may be substituted for sugar. That is, if a formula calls for two pounds of sugar, replace it with one pound of honey. One pound of honey can be substituted for one pound of molasses. Molasses and honey should only be used in the place of or in addition to sugar when the temperature during curing can be kept between 38 and 40° F. Warmer temperatures may make your brine ropy.

Before you decide what cure you are going to use, consider what each ingredient does. Salt preserves the meat; there's no way to get around using it. Sugar is generally added to tenderize and flavor the finished product, but it plays no actual part in preserving the meat. Saltpeter is added primarily to preserve the color of pork and beef. It is not an essential ingredient, if you don't mind your meats a little darker in color. In other words, you can get away with just using salt. As a matter of fact, many old-timers did.

Before curing, make sure that all your meat is chilled to 38 to 40° F. Keep it at this temperature during curing. Don't let it freeze, because the cure won't work when the meat is frozen. If the meat does happen to freeze, extend the curing time for as many days as the meat is frozen. For example, if the curing time is twenty-eight days, and the meat freezes for two days, cure the meat for a total of thirty days. Follow the suggested times carefully until you develop a feel for curing. Never alter curing times until you know what you are doing. Meat that is underexposed to a cure will have a poor keeping quality, whereas meat that is kept in cure too long will shrink excessively and be too salty.

Weigh the meat and curing ingredients. It's very important to use the right kind of salt. Never use iodized salt. Good salts are marketed under such names as rock salt (the kind you use for freezing homemade ice cream and melting ice on sidewalks), dairy salt, or water softener salt. It might be necessary to crush or grind the crystals, since a medium-grind salt is best for curing. A small grain mill, like a Corona or a Quaker City mill, will do a good job of grinding the crystals.

Dry Curing

Dry curing involves rubbing the cure on the outside surfaces of the meat and allowing it time to penetrate. Remember, even though we give the original formulas for cures, it is possible to produce a good cured product without using either the saltpeter or sugar.

The U.S. Department of Agriculture recommends the following dry cure formula for 100 pounds of ham or shoulder:

8 lbs. granulated salt
2 lbs. white or light brown sugar or syrup
2 oz. saltpeter

Use one ounce of this mix per pound of ham and three-fourths to one ounce per pound of bacon. Rub this mixture into the meat to be cured. Rub the required amount of cure into bacons all at once; rub shoulders twice, with one-half the amount each time; and rub hams with one-third the amount three times. Resalt hams and shoulders six to eight days after the last salting. Be sure that the mixture covers the surface; avoid breaking outside membranes or meat will become hard and dry during aging. Pack some of this mixture in the shank ends to make sure that this part gets well cured.

Place the meat in a clean dry container. If you are curing meat on open shelves, cover them with plastic so that the cure does not drip on meats resting on shelves below. Make sure meats don't rest in the brine that will result as the moisture from the meat is drawn out by the salt.

It takes seven days for the cure to penetrate an inch of solid meat; therefore if a piece of bacon is two inches thick, cure it fourteen days; a ham that is five inches thick through the aitchbone should be cured thirty-five days. The cured meat may then be smoked, if desired.

If you measure the amount of cure, the meat won't get too salty even if it is left in the brine too long because the correct amount of curing mix per each pound of meat was used.

We found another method of dry curing, using salt only, in a book written a century ago: *Dr. Chase's Family Physician, Farrier, Bee-Keeper and Second Receipt Book* A. W. Chase (R.A. Beal, Ann Arbor, 1873):

> After cutting out the Hams, they are looped by cutting through the skin so as to hang in the Smoke-Room, shank downwards; then take any clean cask of proper dimensions, which is not necessarily to be watertight.
>
> Cover the bottom with coarse salt, rub the Hams with fine salt, especially about the bony parts; and pack them in the cask, rind down, shank to the center, covering each tier with fine salt one-half inch thick; then lay others on them letting the shank dip considerably, placing salt in all cases between each Ham as they are put in, and between the Hams and the sides of the cask; and so on, putting salt in each layer as before directed; giving the thick part of the Ham the largest share. As the shank begins, more and more, to incline downward, and if this incline gets too great, put in a piece of pork as a check. I let them lie five weeks, if of ordinary size, if large, six weeks, then smoke them . . .

BRINE CURING

Brine curing means soaking the meat in a mixture of salt and water. A brine that also contains sugar is called a sweet pickle. There are several methods for testing the strength of the brine. The old way was to add salt to water until there was enough salt in the water to float an egg or potato. The modern method is to use a salimeter, which will measure the strength of the brine very accurately and allow you to make a brine as strong or as weak as you like. The degree of salinity in a brine or pickle ranges from 60 to 95°. The higher the degree, the greater the concentration of salt.

The standard brine formula for each 100 pounds of meat is:

8	pounds salt	2	ounces saltpeter
2	pounds sugar	4 – 6	gallons water

Once the brine is made, the chilled meat is packed into a vat or watertight barrel. To retain the regular shape of cuts of bacon, line up the bacons evenly, against the sides of the container. Stand them upright with the rinds against the container walls. Add enough brine to cover the meat. Four gallons of brine will cover 100 pounds of closely packed meat. More brine will be needed if the meat is packed loosely. The meat must be totally covered by the brine. If pieces float to the surface, which they probably will, push them down and keep them under the brine by placing over them a plate, piece of wood, or something similar, weighted down with a stone, brick, or other heavy object.

Curing time varies with the degree of salinity. An 85° pickle requires nine days of curing for each inch of meat thickness, a 75° pickle requires eleven days per inch, and a 60° pickle requires thirteen days for each inch. After curing the meat may be smoked, if desired.

In the brine cure, it is extremely important that all parts of the meat be exposed to the brine. Special care should be taken if the brine container is packed tightly. Stir the contents every third day; or remove all the meat once a week, pour off the brine, and repack the meat, pouring the same brine, which has been stirred well, over it. The stirring counteracts the tendency for the brine to separate and become stronger at the bottom as the heavier ingredients settle out. This process is called overhauling.

There is a chance that the brine may become sour (or ropy, as it is sometimes called). This results when the temperature of the room is too high. If the brine has an objectionable odor or adheres to your fingers in long strings when you dip your hand in it, change the brine immediately, wash the meat well, and add new brine. Throw out the old brine.

Some meats, such as bacon, will cure faster than larger pieces, such as hams. Remove the smaller pieces from the brine when they have been cured for the required time and hold them in a cold place until the rest of the meat is finished curing.

Pumping Your Meat

A variation on brining is injecting some of the brine directly into the meat tissues with the aid of a brine pump. The pump, which looks like a giant hypodermic needle, injects the brine into the meat, especially around the bones where spoilage is most likely to occur. The meat is still allowed to soak in brine, but the pump is used to assure penetration to and around the bone.

To speed up the brining process, you can pump brine directly into the meat tissues with a brine pump. It is still necessary to soak pumped meat in brine, but the curing time is shortened because brine has already penetrated the meat.

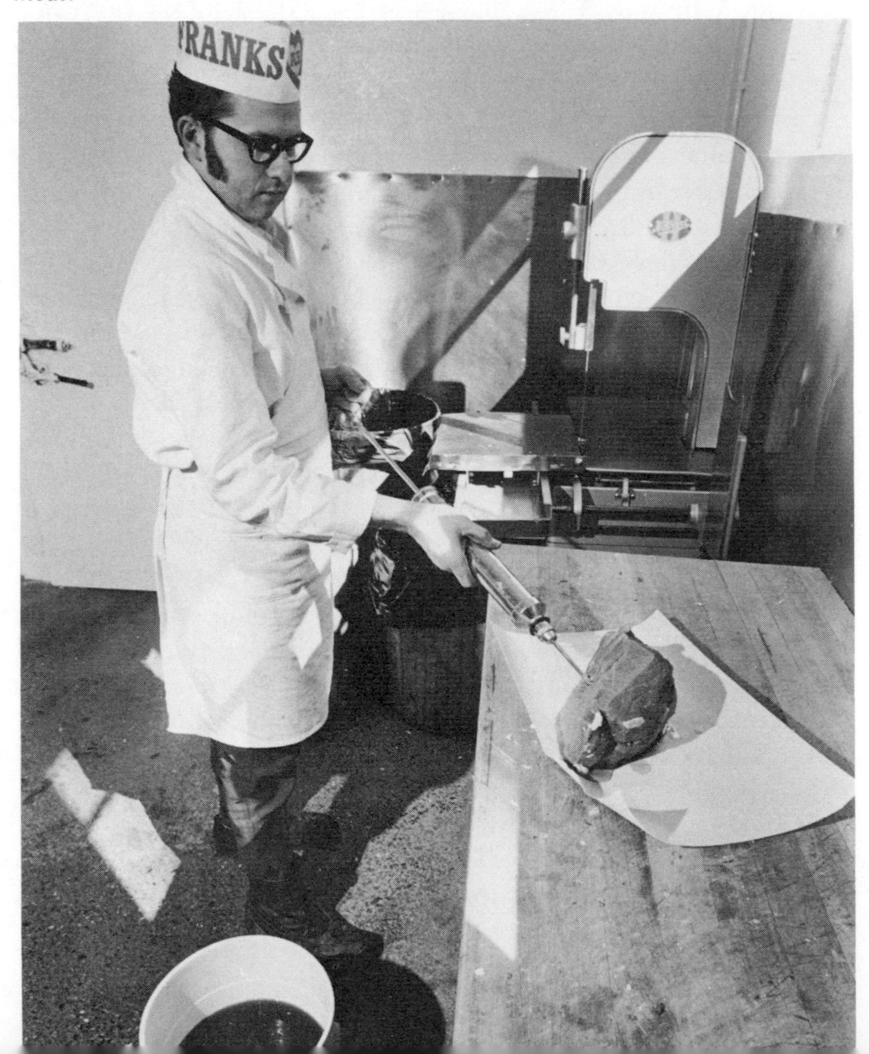

If the meat is to be pumped, it should be chilled and given about four guns, or cylinders, of the brine mixture. The brine is generally a little stronger, perhaps 5 to 10° stronger, when it is to be pumped. When pumping, drive the needle into the thickest part of the meat, slowly, but with steady pressure. Insert the needle close to the bone if there is a bone. Release the brine slowly as you plunge in and also as you draw the needle out so as to distribute the brine evenly around the path of the needle. Remove the needle when it is empty and hold your finger over the hole that the needle has made for a few seconds to prevent any brine from oozing out of the meat. Repeat this procedure in the thickest parts with the remaining cylinders. Curing time is shortened almost one-third for pumped meats because the brine has already penetrated the meat.

Combination Dry and Brine Cure

P. Thomas Ziegler, in *The Meat We Eat* (Interstate Printers and Publishers, Inc.) recommends giving hams cured on the farm, where temperature during curing and curing time may vary, a combination dry cure and brine or sweet pickle. This method hastens salt penetration, and the quicker the salt gets to the center of the ham, the less danger there is of loss from spoilage. Experiments have shown that where one and one-half to two pounds of salt for each 100 pounds of pork is rubbed into hams twenty-four to forty-eight hours before they are placed in a 75° brine, they will cure in nine days per inch thickness of meat. According to Ziegler, this one rubbing of salt is all absorbed overnight and has more rapid penetrating qualities because it is not mixed with other ingredients or dissolved in water.

Curing Beef

Up to this point we have limited our discussion of curing meats to pork because pork is generally the only type of meat that is popularly cured and smoked. However, certain cuts of beef, usually the cheaper, tougher cuts, may be cured for corned beef.

To prepare meat for corning, choose well-chilled meat. Cuts usually chosen for corned beef include the plate, rump, chuck and brisket. Remove the bones to make packing easier. Cut the meat into pieces of uniform size. Weigh all the pieces, and allow eight to ten pounds of salt for each 100 pounds of beef. Before packing, sprinkle a layer of salt on the bottom of a clean, watertight container. Place a layer of meat in the container, then more salt and then another layer of meat. Continue in this manner until all the meat is packed or the container is full. Cover the top layer with a good amount of salt.

Allow the packed meat to remain in the salt for twenty-four hours and then cover the meat and salt with a brine solution, made of:

4 pounds sugar
4 ounces saltpeter
2 ounces baking soda
4 gallons water

This solution is sufficient for 100 pounds of meat. Be certain that the meat is completely covered by the brine solution. If the meat is packed loosely, it may require more brine mixture. Readjust the above formula to make the extra amount of brine needed to cover the meat. To keep the meat totally immersed in the brine, push down with a clean board or similar object, and weight this board down with a rock, stone, or other heavy object.

The meat should be allowed to cure for thirty to forty days at temperatures of between 38 and 40° F. Brine that is permitted to get warmer than 40° may become ropy. Check the brine frequently if curing temperature is difficult to control. Dip your fingers into the brine. If the solution sticks to your fingers, forming strands, it has become ropy. Remove the meat from the brine and wash it. Throw out the old brine and wash the container thoroughly with boiling water. Repack the meat and pour newly made brine over it.

At the end of the curing period, remove meat from brine; wash, and drain before wrapping or smoking.

SMOKING

Hanging Your Meat

Cuts of meat that are to be smoked should have a strong cord strung through them so that they will be easy to hang. Hams and shoulders should be strung through the shanks. If you don't have a regular stringing needle, make a hole in the shank with a boning knife and run a strong cord through the opening. In order to have your bacon hang straight and smooth without wrinkles (wrinkled bacon is unattractive and difficult to cut), it is best to insert a skewer through the flank end of the bacon strip and insert two cords, one on each side of the center of the bacon, just below the skewer.

When hanging meat in the smokehouse, make certain that there is space on all sides of each piece for smoke and hot air to circulate freely. No two pieces should be touching one another.

Once the meat is strung, it should be scrubbed with water and a stiff brush to remove excess surface salt and grease. Rinse the meat and allow it to dry overnight before smoking. If not dry before smoking, the meat will come out of the smokehouse streaky.

Hang the meat in the smokehouse so that no two pieces are touching one another. If you're hanging your meat in two levels, hang it so that the top pieces will not drip on the ones below and make them streaky. If the fire is inside your smokehouse, make sure that the meat is at least six feet from the fire. Hang bacon and sausage further from the source of smoke and watch them carefully to prevent them from being oversmoked.

Building the Fire

Build a fire using green hardwoods, corncobs, or hardwood sawdust and shavings. Don't start your fire with paper, sawdust, or shavings. The ashes from these are very light and may rise and adhere to the hanging meat. However, sawdust or wood shavings may be added to cool the fire once it has started. As we said before, never use softwoods, like pine. The U.S. Department of Agriculture tells us that the Indians made an ideal smokehouse fire by radiating branches and sticks from a central point as if making spokes in a wheel. This arrangement produces a low, cool fire. If the fire gets too hot, or flames rather than smokes, smother it with damp leaves or sawdust. Do not allow the fire to get too hot. The ideal smokehouse temperature is between 80 and 90° F. This is called a cool smoke and is used to smoke meat that is to be aged or held for many months. This smoking and gradual drying process helps to preserve the meat and give it a mellow smoked flavor.

There is another method of smoking meats—with a hot fire. Hot smoking is done when meats are to be eaten immediately or soon after they have been smoked because this method does not preserve them. No brine cure is involved. This method cooks the meat and gives it a smoky flavor at the same time. Poultry and fish are often smoked at high temperatures this way. We will not discuss this method here because we are primarily interested

The ideal smokehouse fire is a cool, smokey one. Such a fire will dry meat slowly and give it a mellow, smoked flavor.

in smoking meats for preservation, not for merely obtaining a smoky flavor.

How Long to Smoke Meats

Old-timers held their meat in the smokehouse for long periods of time. This long smoking considerably reduced moisture in the meat and provided for extended periods of keeping. We prefer our meats much moister now. A good average smoking time is two days, or five to six hours each day on six consecutive days. Small pieces of meat, like sausages and bacons, must be watched carefully to avoid oversmoking. Color is one guide to length of smoking time. Meat should be amber to chestnut brown in color. You'll have to be the judge here because a lot depends on your tastes, and also on the length of time you intend to keep your meat.

The agricultural extension service at the University of Minnesota makes a suggestion to those people who are using smoking as a means of preserving their meat. They suggest smoking meats lightly at first and leaving them to hang in the smokehouse. Pieces are pulled out and eaten as they are needed. As the season progresses, smoking is repeated, a little at a time, until all the meat is used. This method dries off mold and moisture collecting on stored meat. In addition, a permanent smokehouse is a good place to store smoked meats, providing it is fly-, vermin- and thief-proof.

STORING YOUR MEAT

After smoking, cool the meat and sprinkle it with pepper. Pack it in parchment paper that is thick enough to absorb any grease that may run from the meat, and store it in tightly woven cotton bags. Don't use airtight bags; without some air circulation the meat will spoil. Crumpled newspaper packed around the meat will keep the meat away from the sides of the bags and help to prevent insect attack. Tie the bags securely to keep out insects, making sure that all cords, paper, etc., inside the bags are not sticking out and providing a convenient pathway into the bags for insects. Hang the bags in a cool (45 to 55° F. is ideal), well-ventilated, dry room. Examine the meat frequently for insects. If fly larvae infest the stored meat, cut out and burn all the infested parts. The uninfested part, which is safe to eat, should be put into the refrigerator and eaten as soon as possible.

The U.S. Department of Agriculture offers a suggestion for protecting your cured meats from insect attacks. They recommend burying the meat in a box or barrel filled with cottonseed hulls or in a bin of oats or other grains. If you are using a box or barrel, make sure that it is large enough to allow three or four inches on all sides of the hams you want to bury. Fill the container three or four inches deep with cottonseed hulls, place the ham in the middle on the hulls and cover with hulls to a depth of at least three inches. Inspect the ham about once a month for infestation. If the hulls do become infested with grain beetles, replace the old hulls with clean, new ones. If you are burying the meat in a grain bin, attach a string to one end of the meat and run this string to the surface of the grain and attach a tag to it, so that you can easily locate the meat when need be.

If you notice a sour smell when the meat is removed from the smokehouse, don't be alarmed. This may last for a week or two after smoking, but does not mean that the meat is spoiled. Test the soundness of the meat with a ham trier, or make a similar instrument by taking a ten-inch piece of stiff wire and sharpening one end of it. For hams, run the trier along the bone to the center of the ham from both hock and loin ends. Try shoulders in the shank, at the shoulder point and under the bladebone. If the trier brings out a sweet odor, the meat is sound. If there is an unpleasant odor, cut open the piece and examine if for spoilage. If there is a definte odor of putrefaction, destroy the entire piece.

Mold on the surface of the meat may give it a moldy flavor, but this does not necessarily indicate spoilage. Most mold and mold flavor can be scrubbed or trimmed off. Smoked meat may be brushed with vegetable oil or lard every month to delay mold growth.

Ham may be kept a year or more, but after a year, the hams tend to get hard, the fat may turn rancid, and the flavor gets more pronounced. The exceptions are Virginia and Westphalia hams which are best eaten after one to two years' storage. Bacons do not keep as well as hams and shoulders and should be cooked and eaten soon after they are cured and smoked. Cured meat may be frozen, but for best flavor, do not keep it frozen for more than a month or two. The salt and the nitrates found in saltpeter, when used in curing, increase the meat's ability to absorb oxygen and cause it to become rancid quicker. For freezing cured meats, we refer you to the previous section in this book on freezing meats.

All meats that are cured and/or smoked in the manners described in this chapter must be thoroughly cooked before they are consumed. Curing and smoking are not substitutes for cooking and do not destroy any trichinal larvae (which cause trichinosis that might be present.

SUGGESTED READINGS FOR PREPARING AND PROCESSING MEAT FOR STORAGE

"A Complete Guide for Home Meat Curing," Chicago, Illinois: Morton Salt Company.

Ashbrook, Frank G., *Butchering, Processing and Preservation of Meat* New York: D. Van Nostrand Co., Inc., 1955.

"Home Processing of Poultry," University Park, Pennsylvania: The Pennsylvania State University, College of Agriculture Extension Service.

Levie, Albert, *The Meat Handbook.* Westport, Connecticut: The Avi Publishing Co., Inc., 1970.

Reynolds, Phyllis C. *The Complete Book of Meat.* New York: M. Barrows and Co., 1963.

Wanderstock, J.J. and G.H. Wellington, "Let's Cut Meat," Ithaca, New York: New York State College of Agriculture at Cornell University.

NUTS, SEEDS, AND GRAINS

NUTS

HARVESTING THE NUTS

Nuts should be allowed to fully mature and fall from the tree naturally. Some growers, because they are impatient or want to make the harvesting easier, shake the trees to get the nuts to fall. Shake the tree if you wish, but shake gently. Harsh jolts may cause immature nuts to fall, and rough handling can damage the branches. Whatever method you use to get the nuts to the ground, be sure to gather them as soon as possible. Nuts left on the ground mold quickly, especially in damp weather. This is particularly true with the Persian walnut (also known as English or Chinese walnut) and chestnut. If you don't gather and hull the nuts of the black walnut promptly, the bitter fluid secreted in their hulls penetrates to the kernels and this discolors them and impairs their flavor. The nuts of the Persian walnut, if infested with husk maggots, will not open on the tree. Rather, the nuts will fall to the ground with the hulls still on them. In such cases, Persian walnuts should be picked from the ground soon after they fall and hulled like the black walnuts (see below) before the dark, bitter fluid of the hulls penetrates the nut meats.

HULLING THE NUTS

Black walnuts should be hulled before the hull turns black to secure a light-colored, high quality nut. The importance of hulling was proven to Spencer B. Chase, of the Northern Nut Growers Association, after the experiment he performed one year at harvest time. Nuts from ten trees were collected, and the first batch was hulled within a few weeks after maturity. The remainder was hulled at weekly intervals. These hulled nuts were then stored and cured, and subsequently cracked open for inspection from the following January through March. "Without exception," says Mr. Chase, "the first batch of nuts which had been hulled within a week after maturity were light in color, mild in flavor, and could be eaten out of the hand like peanuts. This was not the case with the nuts hulled later at weekly intervals. These nuts produced "darker kernels and suffered considerably in flavor and overall quality." Most commercial walnut operations dip their nuts in a bleaching solution to get the light color which can be obtained naturally just by hulling the nuts as soon as they are mature.

There are several different ways of getting the tough but porous hulls off black walnuts. The old-fashioned corn sheller does a first-rate job when it is equipped with a flywheel and pulley and is driven by a one-quarter horsepower motor. For small quantities of nuts—a few bushels or so—you can spread the nuts on a hard dirt or concrete road and drive over them with a car or truck until all the hulls are mashed. The trouble with this method is that the nuts tend to shoot out from under the wheels when the car rolls forward to crush them. Be ready to hunt for stray nuts and line them up again under the wheels. You can prevent nuts from shooting out from under the wheel by placing a small quantity of nuts in a wooden trough that is just the width of the wheel. Remove the side from one end of the trough so that the wheel can roll in and out of it. Roll the wheel only slightly when crushing the nuts so that you do not run over the other side of the trough.

Whatever method you follow to remove the hulls, it's best to wear gloves to protect your hands from the acrid fluid secreted from the hulls that will leave persistent stains on your skin. It is this staining agent which also penetrates the hulls and stains the delicate kernels and otherwise impairs their quality and flavor. After hulling, the nuts which are still in their inner shells should be washed with a hose or placed in a tub and rinsed thoroughly.

Not all nut varieties, however, will give you as much trouble as the black walnuts. The hulls of pecans, hickory nuts, Persian walnuts (if not infested with husk maggot), filberts, and chestnuts obligingly open on the tree and let the nuts fall to the ground, so harvesting is merely a matter of picking the nuts off the ground.

DRYING THE NUTS

Once gathered (and hulled and washed, if need be), nuts should be placed in water, and the rotten and diseased nuts which float to the top should be discarded. The nuts must then be dried or cured. Green, uncured nuts are bitter and unpalatable. To dry the nuts, spread them out rather thinly on a dry, clean surface and allow them to dry gradually by exposure to a gentle, but steady, movement of air. A clean, cool, darkened, well-ventilated attic or porch is ideal. Nuts dried this way will not be attacked by fungus or mildew. When done properly, such drying will make for light-flavored, light-colored nut kernels. The nuts may also be spread thinly on wire trays or window screens. Nuts should not be more than two nuts deep on such trays and screens. Deep wire baskets are used by some nut growers, but growers are careful to pour the nuts every four days from one basket into another to prevent mildew from forming.

Drying time varies with nut variety. All nuts except chestnuts contain a great amount of oil. This oil prevents nuts from drying out completely and from becoming hard and brittle. Chestnuts, contain little oil, but much water and carbohydrates. Because of their high water and carbohydrate contents, chestnuts dry out easily and become hard and inedible. Dry chestnuts only for three to seven days. All other nuts need several weeks to dry properly.

Walnuts, pecans, filberts, and hickory nuts are dry enough to be stored when the kernels shake freely in their shells, or when the kernels break with a sharp snap when bent between the fingers or bitten with the teeth. Avoid excessive drying which will cause the nut shells to crack. After drying, nuts still in their shells may be stored in attics for up to a year, but cool underground cellars are preferable for longer storage. Storage containers for nuts may be made from large plastic bags with ventilation holes punched in them to allow air to circulate and excess moisture to escape. These bags are inserted in tin cans lined with paper. They are kept tightly closed, but a small hole should be punched in the can below the lid. Store the nuts at 34 to 40° F. and check regularly for mold and mildew. Some nut growers store their nuts in peat moss and other moisture-absorbing materials. By so doing, there is no need to allow for ventilation because the peat moss takes care of excess moisture that would cause mildew. Prepare peat moss for nut storage by moistening it with just enough water to prevent its being dusty. The peat moss should be damp, but not wet. Add about one-third as much peat moss as nuts by volume when packing nuts in lidded cans or plastic bags.

Nuts may also be packed in clean, dry sand and then stored in a cool area. Stored this way, nuts will retain their germinating powers, but may lose some of their flavor.

CRACKING THE NUTS

You can crack a nut with almost anything that does the job. You can use a hammer with a block of wood or metal. Or you can swing a heavy iron like your grandmother used to iron shirts. Every nut fan has his own method. Take wild foods gourmet Euell Gibbons, for example. Here are his instructions for shelling black walnuts, especially hard nuts to crack:

> If you stand the nut—pointed end up—on a solid surface and hit it with a sharp blow with a hammer, it will crack into two halves. Stand each half, again pointed end up, and strike it again, which will break it into quarters. Strike each quarter again on the pointed end, and it well break into eighths—at which point the nut meats fall out.
>
> When the nuts are well dried, I have been able to completely empty shell after shell without resorting to a nutpick with this method.

And his secret for shelling hickory nuts:

> Put the unshelled nuts in the deep-freeze for a day, then remove a few handfuls at a time and shell them while they are still frozen and brittle. The nuts are narrower one way than the other, so hold them up edgewise and strike each a sharp blow with the hammer, just hard enough to crack the shell but not smash the kernel. Cracked this way, many entire halves can be removed while the others usually have to come out in quarters.

Hard-shelled nuts are more easily shelled if the shells are softened first. This may be done by pouring boiling water over the nuts and allowing them to stand for fifteen to twenty minutes before shelling. Another method of softening the nuts is to place them in a container, sprinkle them with water, and cover them first with a damp cloth and then the container cover. Let them stand this way for twelve to twenty-four hours. The shells will be easier to crack, and the nut meats will not splinter when the shells are cracked.

Soft-shelled nuts, like chestnuts, can have their shells peeled off with a knife. To do so, first cut across the base of the nut and then place the nut in boiling water for three to four minutes. The shell will then pull off easily.

STORING SHELLED NUTS

Shelled nuts don't store as well as those with their shells intact. Without their protective shells, nut meats are exposed to light, heat, moisture, and air—all of which cause rancidity in the nut. If you plan to keep your nuts for a while, crack them as you need them. Nut meats can be stored safely for a few months under refrigeration if they are placed in tight-closing jars or sealed plastic bags.

For longer keeping, nut meats can be frozen or canned. They will keep well at freezer temperatures for up to two years so long as they are stored in airtight containers or plastic freezer bags. Sealed in jars, nuts meats will keep well for one year.

Nut meats to be canned should be spread in a thin layer on a cookie sheet or baking pan and heated for thirty minutes in an oven preheated to 250° F. Stir the nuts every ten minutes to make sure that they are drying evenly. Fill dry, sterilized canning jars with the hot nut meats and process them in either a boiling-water bath or a steam-pressure process:

For a boiling-water bath:

1. Pack hot nut meats in canning jars and screw lids on tightly.
2. Place the jars on rack in boiling-water canner so that no jars are touching one another. If you are placing jars two layers deep, separate the layers with a metal rack and be certain that no jar is directly on top of another.
3. Have enough hot water in the canner to come up two inches over the top of the jars. Put the lid on the canner.
4. After the water in the canner comes to a boil, process for thirty minutes. Boil gently and steadily, adding more boiling

water if needed to keep jars covered. Do not pour the water directly over the tops of the jars.

5. As soon as thirty minutes are up, remove the jars from the canner.
6. Complete the lids on the jars as soon as they are removed from the canner, if necessary.
7. Set jars upright to cool. Place them far enough apart from one another so that air can circulate freely around them.

For steam-pressure processing:

1. Pack hot nut meats in canning jars and screw lids on tightly.
2. Place the jars on rack in the canner, making sure that no two cans are touching. If your jars are two layers deep, place a metal rack between the layers and be certain that no jar is directly over another.
3. Put two or three inches of hot water in the bottom of the canner.
4. Fasten the canner lid securely so that no steam escapes except at the open petcock or weighted gauge opening. Then close the petcock or weighted gauge and let the pressure rise to five pounds.
5. Process for ten minutes.
6. After ten minutes, gently remove the canner from the heat and let the canner stand until the pressure returns to zero. Wait a minute or two and slowly open the petcock or remove the weighted gauge.
7. Remove the jars and complete the seals if necessary. Allow the jars to cool.

Label the jars and store them with your other canned foods, in a cool, dry, dark place.

RECIPES

Roasted and Salted Nuts

Remove the nuts from their shells and skin them, if you wish. To make skinning them easier, place nuts in a bowl and pour boiling water over them. Leave them in the boiling water for about 3 minutes, until the skins begin to wrinkle. Pour the water off the nuts and rinse them under cold water to cool them. Now, rub the skins off with your fingers; they should come off easily.

Place the nuts in a bowl, and for each cup of nuts, sprinkle one teaspoon of melted butter or oil (peanut, sesame, or safflower oil) over them. Stir the nuts so that all are coated with oil. Then spread them out on a cookie sheet, sprinkle lightly with salt, and roast in a 350° F. oven for about 10 minutes, until the nuts are just lightly browned. Watch the nuts carefully in the last few minutes of roasting so that they don't become too dark.

Dry Roasted Nuts

Skin the nuts, if you wish, as described above. Place the nuts in a bowl, and for each cup of nuts, pour one teaspoon of soy sauce over them. Mix the nuts so that all are coated with the soy sauce. Spread on a cookie sheet and bake in a 350° F. oven for about 10 minutes, until the nuts are lightly roasted. Watch the nuts carefully in the last few minutes of roasting so that they don't become too dark.

Cucumber, Nuts, and Olives

2 cucumbers chopped
2 cooked eggs
1 cup stuffed olives chopped
½ cup chopped pecans or almonds
½ cup chopped celery
yogurt

Toss all chopped ingredients with yogurt or salad dressing. This is good on rice wafers, rye crisp, or turnip slices. Yield: 4 servings

Cabbage-Rice Casserole

1 head of cabbage steamed and chopped
1 cup cooked brown rice
1 tablespoon chopped parsley
2 tablespoons apple cider
1 teaspoon fennel seed
½ cup chopped nuts

Oil a baking dish, line the bottom with half of the chopped cabbage. Lay the cooked rice, nuts and seasoning over the cabbage, moisten with 2 tablespoons of apple cider. Add another layer of cabbage, then rice and nuts and the cider. Lay several coarsely chopped apples on top and bake for about 30 minutes at 325° F. Yield: 4 servings

Walnut Loaf

Cover 1½ cups of wheat germ with milk and let the wheat germ absorb all the liquid it will take up. Drain slightly. Place in mixing bowl and mix with:

1 cup chopped walnuts
1 chopped onion
1 chopped green pepper
1 chopped tomato
juice of one lemon
1 tablespoon oil
1 egg, well beaten

Form into a loaf, place in oiled pan and bake at 325° F. about 30 minutes. Serve with tomato or mushroom sauce. Yield: 4 servings

Waldorf Salad

1 cup chopped apples
1 cup English walnut halves
½ cup plumped raisins
½ cup chopped sunflower seeds

Blend, add dressing and serve on salad greens. Yield: 4 servings

Chestnut Stuffing

3 cups chestnuts
3 cups diced celery
3 cups wholewheat bread-crumbs
⅓ cup cream
¼ teaspoon mace
1 teaspoon salt
2 tablespoons butter

Boil the chestnuts, peel, and blanch. Break into small pieces. Brown the celery in the butter, add the chestnuts and the crumbs, and mix. Moisten with the cream, sprinkle with salt, and stuff loosely.

Nutty-Rice Loaf

1 cup chopped nuts
1 cup cooked brown rice
1 cup wheat germ
2 eggs, beaten
2 onions, chopped
¼ cup celery and tops, chopped
½ green pepper, chopped
1 cup water
½ cup soy flour
½ teaspoon sea salt
½ teaspoon sage (or 2 tea-spoon fresh sage)
a few grates of fresh pepper-corns

Blend all ingredients, and turn into a greased loaf pan. Bake at 350 degrees for 45 minutes. Serve with tomato sauce or your favorite sauce. Yield: 6 servings

Eggplant-Walnut Delight

1 medium sized eggplant
1 medium sized onion
½ cup walnut meats
sea salt and oregano to taste
½ cup tomato sauce

Peel eggplant and cut into chunks. Put eggplant, onion and nut meats through a food grinder. Add seasonings and tomato sauce and mix thoroughly. Form into a loaf and bake in an oiled casserole for 20 minutes. Serve hot. Yield: 4 servings

Almond Loaf

Pour 1 cup of hot stock or potato water over 1 cup of oatmeal, blend together, cover and let stand until cool. Add the following:

2 eggs
1½ cups ground almonds
½ cup ground sunflower seeds
1 medium onion, chopped
½ cup diced celery and tops
½ chopped bell pepper or pimiento
1 teaspoon marjoram (fresh or dried)
1 tablespoon sea kelp
1 tablespoon oil

Blend and add as much more of the stock or potato water as is needed for loaf consistency. Pour into an oiled bread tin and bake from 30 to 45 minutes, or until done, at about 350 degrees. Turn out on a meat platter and garnish with red pepper and parsley. Serve hot or cold. Yield 4 servings

Nutty Burgers

1 cup nuts, ground fine
½ cup sunflower seeds, ground fine
½ cup grated carrots
1 stalk celery, ground
1 small onion, ground
½ green pepper, ground
1 sprig parsley, ground

Put the above ingredients through the food grinder or in the blender with enough liquid (a tomato or ½ cup broth) and then add the following:

2 eggs
1 teaspoon sea kelp
1 pinch of sage or marjoram

Blend the ingredients and if they are too dry, add broth or tomato pulp; if too moist, add rice polishings. Shape into patties and broil until browned on each side. Yield: 4 servings

Nut-Coconut Pie Shell (Unbaked)

½ cup nuts, ground
½ cup shredded coconut
⅓ cup oil

Blend all ingredients. Press into 9-inch pie pan. Chill and fill.

Chewy Squares

1 egg, separated
⅓ cup honey
⅛ teaspoon salt
½ cup nuts, chopped

3 tablespoons wholewheat pastry flour
¼ cup ground sunflower seeds

Beat egg yolk until thick. Blend in honey. Combine with flour, salt, nuts, and seeds. Fold in stiffly beaten egg white. Turn into oiled square pan. Bake at 350° F. for 20 to 25 minutes, until light brown. Cool. Cut into squares. Yield: 9 squares

Coconut-Nut Balls

2 cups wholewheat pastry flour
3 tablespoons oat flour
¼ teaspoon salt
4 tablespoons honey
1 cup oil
3 tablespoons coconut
2 cups nuts, ground coconut

Sift flours and salt. Blend in honey and oil. Add nuts and coconut and mix well. Shape into ½-inch balls. Place on lightly oiled cookie sheet. Bake at 350° F. about 30 minutes. When cool, roll in coconut. Yield: 3 dozen ½-inch balls

Walnut Torte

6 eggs
¾ cup honey
½ cup powdered skim milk
½ cup wheat germ
2 cups walnuts (ground fine)
1 teaspoon vanilla extract

1. Preheat oven to 325° F. Prepare two 9-inch layer cake pans by oiling bottoms of pans with pastry brush. Cut two circles from heavy brown paper. Place a circle of heavy brown paper on the bottom of each pan. Brush the paper thoroughly with oil.
2. Separate eggs, putting whites into large bowl of electric mixer and yolks into smaller bowl reserving two yolks for filling.
3. With mixer at high speed, beat whites until stiff peaks form when beater is slowly raised. Set to one side. With same beater, beat yolks until thick and lemon-colored. Gradually blend the honey into the yolks.
4. Stir in the powdered milk, wheat germ, and ground walnuts. Blend together. Add vanilla extract.
5. With wire whisk or rubber scraper, using an under-and-over motion, gently fold yolk mixture into beaten egg whites until well combined.
6. Pour the batter into the prepared pans, spreading evenly to edge. Bake in a slow oven for 30 minutes.

7. Remove from oven and loosen sides with a spatula to ease the cake out of the pan. Invert pans on a wire rack and remove paper immediately. Cool cakes completely before frosting.
8. Make Custard filling. Cool. Put layers together with filling. Frost top and sides with frosting.
9. This recipe makes 10 to 12 servings.

Ground pecans or ground almonds may be substituted for the walnuts. When using ground almonds, use ½ teaspoon almond extract in torte and frosting in place of the vanilla extract.

Custard Filling:

⅓ cup cornstarch
¼ cup powdered skim milk
¼ teaspoon salt
1¾ cups water
⅓ cup honey
2 egg yolks, slightly beaten
1 teaspoon vanilla extract

1. In medium saucepan, combine cornstarch, powdered milk and salt.
2. Add ¼ cup of water, gradually, stirring with wooden spoon until mixture is smooth and free of lumps. Add remaining 1½ cups of water, mixing thoroughly.
3. Add honey to mixture and place over medium heat, stirring constantly until custard thickens; should take from 10 to 12 minutes.
4. Remove custard from heat. Add three tablespoons of hot mixture to beaten egg yolks. Mix well. Gradually pour yolk mixture into custard, blending well.
5. Return to medium heat and cook three minutes, stirring constantly. Remove from heat. Add vanilla. Cool custard completely before using to fill cake.
6. Makes approximately two cups of custard.

Frosting:

½ cup honey
2 egg whites
½ teaspoon vanilla extract

1. In top of double boiler, combine egg whites and honey. Beat one minute with rotary beater (hand or electric type), to combine ingredients.
2. Cook over rapidly boiling water (water in bottom should not touch top of double boiler), beating constantly until soft peaks

form when beater is slowly raised. Allow about 10 minutes for beating to get proper consistency.
3. Remove from boiling water. Add vanilla. Continue beating until frosting is thick enough to spread—about 4 minutes.
4. Makes enough frosting for a 9-inch two-layer cake.

Nutty Waffles

1¾ cups wholewheat pastry flour
2 teaspoons baking powder
½ teaspoon salt
1 tablespoon honey
3 egg yolks
4 tablespoons melted butter or oil
1½ cups milk
3 egg whites
½ cup chopped nuts

Mix together the flour, baking powder, and salt. In another bowl, beat together the egg yolks, honey, butter or oil, and milk.

Make a well in the dry ingredients and pour in the milk mixture. Combine them with a few, swift strokes. Mix in the nuts.

Beat the egg whites until stiff and fold them into the batter briefly. Pour the batter by the tablespoonful into a hot waffle iron and cook for about 4 minutes or until brown and crispy. Yield: 6 waffles

SUNFLOWER SEEDS

The back of a thoroughly ripe sunflower head is brown and dry, with no trace of green left in it. But, the trouble with leaving it in the field until that point is that the birds may have harvested the crop for you, or the head may have shattered and dropped many of its seeds to the ground.

SMALL-SCALE HARVESTING

In the home garden, sunflower seeds may be covered with cheesecloth to keep away birds. Or the heads may be cut off when the seeds are large enough and allowed to dry elsewhere. If cut with a foot or two of stem attached, they may be hung in a dry, well-ventilated place to finish drying. They may also be cut

A home-improvised arrangement, such as this wire brush attached to a small electric motor, can simplify removing the seeds from the pods—often a difficult and tiring task for gardeners with a large quantity of sunflowers to harvest.

and spread on boards on the ground—protected with a wire screening from rodents—to dry in the sun for a couple of weeks. Heads should not be piled on top of each other, as seeds may become moldy or rot.

The heads are dry and ready to have their seeds removed when the rough stalks are brittle and the seeds separate from the head easily as you run your thumb lightly over the surface of the head. You can brush the seeds out with a stiff brush, a fish-scaler, or a currycomb; or you may remove them by rubbing the heads over a screen of half-inch hardware cloth stretched over a box or barrel. If some seed is still moist it may be spread out to complete drying after being removed from the head.

LARGE-SCALE HARVESTING

For large-scale harvesting the larger-sized varieties of sunflowers, which are the types grown for seeds, present a problem. They are too high and too heavy to be handled by a combine, and the toughness of the heads makes threshing difficult. The smaller sunflower varieties can be combined, but their yields are not big and their seeds are small.

It is best to allow the sunflowers that must be harvested by hand to dry in the fields on the stalk. The birds and animals will primarily pick at those heads that fall to the ground. If the heads are taken indoors to dry, rats may get to them.

The dry heads are harvested by driving around the edge of the field with a wagon or truck and clipping the heads off with pruning shears. The heads are then run through a corn fodder shredder which knocks the heads apart and loosens the seeds. The seeds should then be screened to separate them from pieces of the stalk and the head.

HULLING YOUR SEEDS

For hulling small amounts of sunflower seeds you can make the hand hulling job easier by putting the seeds in boiling water for a short time. You can also crack the shells with pliers or a clothespin. Lay the seed between the prongs as you would between your teeth, so that it will be cracked lengthwise.

To hull large quantities, use any small farm hammermill. A ten-inch size is adequate. Remove screens and set the mill to run at 350 revolutions per minute. You can use a tachometer (which measures the speed of rotation) to judge so no hulled seeds go up the dust collector. Slowly pour in about two gallons of seeds, then speed up the mill to 1,200 to 1,500 revolutions per minute. This clears the mill so that it won't clog at the bottom. Put any container under the spout from the dust collector. Such a mill will hull out about 80 percent or more of the seeds.

Local feed mills may be willing to shell sunflower seeds in amounts too large for hand hulling, if you haven't a hammermill or quantity large enough to make purchasing one practical.

STORING YOUR SEEDS

Dried seed should be stored in a cool, dry place in small containers and should be stirred once or twice a week to prevent mustiness. Seeds that are stored in large bins may heat up and lose some of their vitamin content.

Hulled seeds may also be canned or frozen for long-time keeping. If packed in freezer containers or freezer bags, sunflower seeds will keep for a year or more at freezer temperatures. Can seeds as you would nut meats; see information for canning in the section on harvesting and storing nuts.

Sunflower Goulash

- 1 cup chopped sunflower seeds
- 2 cups mushrooms, halved
- 2 onions, sliced
- 1 cup fresh sprouts
- 1 cup cooked and cooled millet
- 2 cups fresh or home-canned tomatoes
- 1 cup lima or butter beans
- 1 teaspoon chili powder
- sunflower seed oil

Sauté the onions in sunflower seed oil until tender. Add all other ingredients, season, pour in casserole dish and bake about ½ hour, and serve. Yield: 4 servings

Sunflower and Wholewheat Bread

- 1 package yeast
- ⅓ cup lukewarm water
- ¼ cup honey
- 1¼ cups boiling water
- 2 teaspoons salt
- 2 tablespoons butter
- 1¾ cups wholewheat flour
- 1¾ cups sunflower seed flour
- ⅓ cup raisins

Soak yeast in the lukewarm water and sweetening until foamy. Boil water, salt, and butter and then let it cool to lukewarm. Mix the flours and raisins together. Combine the yeast mixture with the water and add it to the flour. The sponge should be stirred well and be soft, but with a body. Set the bowl in a pan of quite warm water and cover with a tea towel to rise. This takes about 1¼ hours. Beat down a couple of minutes with a wooden spoon and then put in a greased, floured bread pan to rise again in a warm place, covered, about 45 minutes. Light the oven when you put the bread in and let it come up to 375° F. and then turn down to 325° F. to finish the baking which takes about 45 minutes altogether.

Sunflower Corn Bread

- 1 cup sunflower seed flour
- 1 cup corn meal
- 2 teaspoons baking powder
- 1 teaspoon salt
- ½ teaspoon soda
- 2 egg yolks, beaten
- ⅓ cup sour cream or yogurt
- 2 tablespoons honey
- ⅓ cup buttermilk
- 2 egg whites, beaten

Mix all dry ingredients together and combine with the mixed liquids. Fold in the whites last. Bake in a greased, floured pan 5 minutes at 375° F. then 10 minutes more at 325° F. Serve hot with butter. It keeps well and is good cold.

Sunflower Seed Pie

Crust—Make your pastry the usual way except that you add ½ cup finely ground sunflower seeds to the shortening and use a little less flour. Bake the shell and cool it.

Filling—This is an old-fashioned cream pie filling. Heat 2 cups milk to boiling point and stir in a thickening of:

⅔ cup honey
2 egg yolks
½ teaspoon salt
3 rounded tablespoons flour (unbleached)

Stir this to a paste with milk. Stir constantly until thick. Cool and add 1 teaspoon vanilla. Pour in pastry shell, smooth over the top and sprinkle about ¾ cups ground sunflower seeds on the filling. Cover with the egg whites made into a meringue and sweetened with honey. Bake to a golden brown on top.

Sunflower Seed Loaf

1 cup ground sunflower seeds
1 cup ground sunflower seeds (seeds ground fine)
½ cup ground or chopped walnuts
1 cup cooked and seasoned lentils
2 tablespoons minced onion or chives
2 large or 3 small eggs beaten slightly
½ cup wholegrain bread crumbs
½ cup grated raw carrots
1 tablespoon oil
½ cup diced green pepper
1 teaspoon sea salt
½ teaspoon paprika
¼ teaspoon thyme
2 teaspoons lemon juice

Blend ingredients and press in a greased baking dish. Bake 45 minutes or until done in the middle at 350° F. and turn out when cooled a bit. Serve with mushroom sauce if you want it still more valuable as a protein dish.

Sunburgers

1 cup ground sunflower seeds
½ cup finely chopped celery
1 tablespoon chopped parsley or pepper grass

2 tablespoons chopped onions or chives
1 egg (not absolutely necessary)
1 tablespoon chopped raw pepper
½ cup grated raw carrots
1 tablespoon oil
½ teaspoon sea salt
¼ cup tomato juice
1 pinch basil

Combine with enough tomato juice so that the patties hold a good formed shape, arrange in a shallow baking dish and bake at about 350° F. until browned, turn and brown on other side. They can also be broiled if coated with oil on both sides before cooking. They can be served with mushrooms, grated cheese, or fixed up in various ways, but they make a perfect protein dish.

Sunflower Seed Bread

One compressed yeast cake dissolved in ⅓ cup warm water. Dissolve ⅓ cup each of honey and shortening in 1½ cups very warm water. Add ½ cup raisins and let cool. Add 1½ cups ground sunflower seeds to cooled mixture then add yeast plus the following ingredients:

2 teaspoons sea salt
1 teaspoon nutmeg
½ cup wheatgerm
1 teaspoon cinnamon
3 cups whole grain bread flour

Stir into moderately stiff dough, grease the top and let rise. Stir down and spoon into 2 small greased bread tins. Let rise again and bake about 50 minutes at about 350° F.

Sunflower Seed Cookies

2 egg whites
1 cup honey (thinned down over hot water)
1 cup chopped sunflower seeds
1 teaspoon vanilla

Beat the egg whites until they will hold a peak, then gradually beat in the honey, then the seeds which may be ground to your particular taste and finally add the vanilla. Drop cookies on a heavily greased baking sheet and bake at 275° F. from 30 to 40 minutes at center of oven. Test and take out when toothpick comes out clean. Loosen at once as they like to stick to cookie sheet. Keep in a container with a tight lid.

Sunflower Seed Muffins

2 tablespoons honey
1 egg
1½ cups wholewheat pastry flour

½ teaspoon salt
2 tablespoons oil or melted butter
1 cup fresh or powdered milk (diluted)
½ cup ground sunflower seeds
2 teaspoons baking powder

Blend egg, honey and oil. Add milk and sifted dry ingredients and pour in heated muffin tins. Bake 20 minutes at 400° F.

Gradually use less baking powder and depend on egg white to leaven the batter each time you make these muffins. This is how it is done. Use 2 eggs, separated. Use the yolks with the batter and beat the whites stiff and fold in just before baking. Use less baking powder, even using 3 egg whites if needed, until you need no baking powder at all.

Sunflower Sour Cream Muffins

¾ cup sunflower seed flour
1 cup of rye, wholewheat or oatmeal flour
2½ teaspoons baking powder
1 teaspoon salt
¼ cup currants
½ teaspoon soda
1 egg beaten
2 tablespoons honey
¾ cup sour cream
⅓ cup buttermilk (about)

Mix all dry ingredients with the currants, then combine with the mixed liquids, stirring as little as possible. Add buttermilk last, as needed, to make a soft dough. Rye takes about ⅓ of a cup, but there is a difference in flours. Wholewheat or oatmeal may take more or less. Bake in a greased muffin tin for 5 minutes at 375° F., then 10 to 13 minutes more at 325° F. Yield: 9 muffins

Sunflower Baking Powder Biscuits

½ cup sunflower seed flour
½ cup wholewheat pastry flour
¾ teaspoon salt
1¾ teaspoons baking powder
2 or 3 tablespoons lard
⅓ cup milk (about)

Sift the dry ingredients, then cut in the lard with a pastry knife. Three tablespoons of lard make a very rich biscuit, two tablespoons make a puffier and a little lighter one. Mix just enough milk in the dough to make it soft but firm. Drop from a spoon onto a greased, floured pan and bake at 375° F. 10 to 12 minutes.

Sunflower Drop Biscuits

½ cup wholewheat flour
½ cup sunflower seed flour
2 teaspoons baking powder
¾ teaspoon salt
¼ cup sour cream or yogurt
3 tablespoons milk (about)

Mix dry ingredients together and cut in the cream with a fork as you would fat. Then moisten with just enough milk to mix the dough. Drop from a spoon on greased, floured pan and bake 8 to 10 minutes at 375° F. Yield: 12 small biscuits

Sunflower Pancakes

¾ cup sunflower seed flour
¾ cup wholewheat flour
1 teaspoon baking powder
¾ teaspoon salt
2 eggs, beaten
1 tablespoon honey
2 tablespoons butter, melted
1½ cups buttermilk

Sift all dry ingredients together. Mix the liquids together and combine with the flour. Fry a delicate brown on both sides.

They are very tender and good with just butter. They would be fine for filling with creamed chicken or sea food and then rolled and kept warm a minute in the oven. They have an unusual flavor which can be enjoyed without syrup. Yield: 20 – 22 pancakes

Sunflower Seed Omelet

4 eggs beaten very light
1 cup sunflower seed meal
½ teaspoon kelp
½ teaspoon caraway seeds

Heat a heavy skillet. Blend ingredients. Oil the hot skillet, pour in the mixture, let brown on bottom, cut in quarters, turn and brown on the other side. Yield: 2 – 3 servings

GRAINS

Knowing that the breads and cereals you eat have been made from organic grains you have grown yourself is truly satisfying. Compared to growing fruits and vegetables, growing grain is almost effortless. What little hard work there is comes at the end, when you must harvest and prepare the grains for storage.

If you were a big operator, you would wait until your grain was dead ripe and had dried down to a moisture content of about

14 to 14½ percent, and then get out the combine that would do the whole job of harvesting: cut the stems, beat the grain loose from the hulls, and separate the straw and chaff from the grains. Then you would run the grain through a mechanical fan seed cleaner to remove the wild plants and seeds that were harvested with the grain. After the grain is cleaned you might machine dry it to prevent it from heating up and spoiling and to retard mold and fungus growth during storage. Then you would pour the cleaned dry grain into 100-pound sacks and store them in an atmospherically controlled warehouse.

HARVESTING BY HAND

A combine costs several thousand dollars, and unless a neighbor has one that can be borrowed, a grower with only a small amount of grain to harvest will have to do it the old-fashioned way and harvest the grains by hand. (The cutting can be done with a sickle-bar on a garden tractor, but the machine running over the heads may cause the grain to shatter onto the ground.)

Harvesting by hand involves cutting the grain before it is completely ripe (contrary to combine harvesting), then binding the stalks into sheaves, arranging the bundles into shocks, and allowing the grain to remain drying in the field until it is ripe. The best way to tell if the wheat is ready to cut is to pull a few heads and shell out the grain in the palms of your hands with a rubbing motion. The wheat should come free from its husks fairly easily, but not too easily. Blow the chaff out of your cupped hand and chew a few grains. They should feel hard when biting into them. If the grain is still milky, wait a few more days. If the grains are very hard and shatter out of the husks very easily, you've waited too long. The stalks should be nearly all yellow, with only a few green streaks remaining. When wheat is dead ripe, there is no more green in the stalk.

Rye should be harvested in the almost-ripe stage, which it usually reaches about a week before the wheat does. Oats are hand harvested when the kernels can be dented by the thumbnail—not too hard and not too soft—when the heads are yellow and some leaves are still green. Buckwheat, which takes longer to mature, should be harvested after the first seeds mature.

Ideally, this is after one or two frosts have made the grains easier to separate from the plants.

CRADLING

To harvest by hand, cut the stems near the ground after the dew is off, on a dry day. To do this properly, according to homesteader, Gene Logsdon, you need a grain cradle. A grain cradle is a scythe equipped with three or four long wooden tines arranged about six inches apart above the scythe blade. When you swing the cradle, the cut stalks of grain gather against the wooden tines as you make your stroke through the standing grain. The cut stalks then fall in a neat little pile to the left of the swath you are cutting as you complete your stroke. These little piles are then easily tied into bundles.

It takes practice to develop the proper rhythm for cradling. But if you've ever done scything, you can catch on to it in a hurry. The trick of scything is to cut a rather narrow strip, letting the scythe blade slice through the standing stalks at a sidewise, 45-degree angle. Don't try to whack off the stalks with the blade at right angles to the stalks. The blade should be very sharp and always held parallel to the ground. Don't let the point dip down to catch the ground.

If a grain cradle isn't available, an ordinary scythe will do. This cuts the stalks, but instead of letting the grain fall in convenient piles, it lets the stalks fall where they are cut. This means that you have to gather the stalks in bundles, which could become a laborious job if you've got a big area to harvest. A hedge trimmer could conceivably be used to cut down the grain, too. Like a scythe, a hedge trimmer would let the stalks fall haphazardly, making gathering difficult. It wouldn't give you the aching arm and back a manually-operated cradle or scythe would, but you would have to consider the practicality of using a power tool in the field. You would have to have an extra long extension cord for an electrically powered trimmer or one that runs on gasoline.

After the stalks are cut, tie them into bundles with ordinary binder twine, available from farm stores. A bundle should measure about eight inches in diameter at the tie. The bundles are then set up in shocks.

BUILDING THE SHOCKS

To build a shock, grab a bundle in each hand, sock the butt ends firmly down on the ground and then lean the tops against each other. Two more bundles are set the same way on either side of the first ones. Sometimes your beginning shocks will fall over until you get the knack of it. With the first four bundles in place, you can then stand about six or eight more evenly around them. When you have a fairly sturdy shock of twelve bundles or so, you can tie a piece of twine around the whole shock to make it stand more solidly. After the shocks dry in the sun for about ten days (a few days longer for oats) put the bundle in an airy shelter to finish ripening where no rain would fall on them.

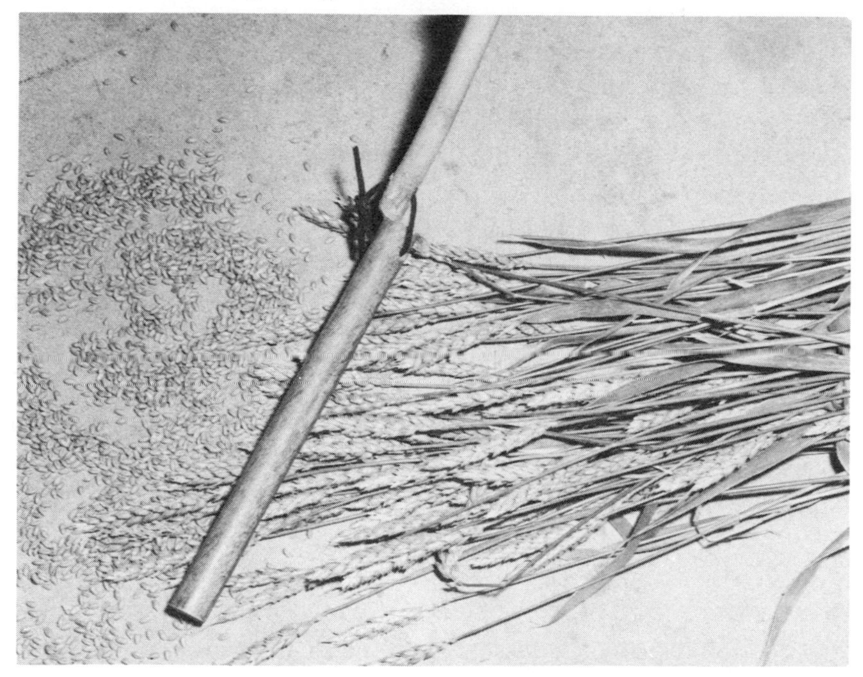

One way to thresh small quantities of grain is to lay a handful on a hard surface and beat the grains with a homemade flail, like the one pictured here.

THRESHING

Before the advent of the combine, a steam-powered threshing machine would make the rounds and farmers would take their sheaves down out of the mow and feed it to the roaring, hungry monster. Before that—for thousands of years, and it's still done this way in many parts of the world—people beat the grain out of the hulls with flails.

You can make a simple flail by taking an old broom handle and drilling a quarter-inch hole near one end. Then similarly drill a three-quarter-inch stick about a foot long and attach loosely to the broom handle with a leather or wire loop through the drilled hole. Throw a sheaf of wheat on the threshing floor—a clean garage or cellar floor or a hard-packed piece of ground free of growth or debris. Then beat the heck out of the sheaf with the loose end of the flail; the grain will easily fall out.

A flail isn't the only instrument you can use for this job, though. Gene Logsdon uses a homemade flail, but recommends trying a plastic baseball bat instead. His son used it and found that it worked better than the flail because it was firm enough to knock the kernels out, but flexible enough not to crack the grains. David Criner in Arkansas uses a compost shredder to thresh his grain. He places a tarpaulin under the shredder to catch the grain and feeds the bundles of wheat—one at a time, head first—into the shredder chute. Another simple way to thresh is to wham a handful of stems against the inside of a barrel. The grain falls neatly into the barrel instead of flying around and landing in a wide area on the floor.

REMOVING THE CHAFF

Once the straw bundles are relieved of their grain and removed, there still remains lots of chaff with the grain. Traditionally, grain was winnowed simply by pouring it slowly into a bucket in a brisk wind. The breeze blew away the light husks, and the heavier grain fell directly to the bucket. If the day you pick is windless, use an electric fan to generate a winnowing wind; it works like a charm. Repeat this process until the grain is clean, or nearly so. A few grains may not hull out completely, but this will not effect the flour.

STORING YOUR GRAIN

Curing

If you harvested your grain when it was almost ripe and allowed it to dry in the field for at least a week or so, the moisture content of your grain is probably somewhere between 12 and 15 percent. If this is the case, your grain is dry enough to store as is, although it should be cured for at least one month before you try to mill it for flour. Green grain heats up and gums up a flour mill, and the flour is difficult to work with. Grain makes much better bread if it is allowed to cure longer than a month.

If the weather should be damp when you harvest the grain, and it rains while the grain is curing in the field, there is a good chance that its moisture content is somewhere above 15 percent. If you store high-moisture grain you are inviting all kinds of trouble. Mold, fungi, beetles, and mites thrive in moist grain. Wet grain should be dried before it is stored. The drier the grain, the less susceptible it is to damage by bacteria, fungi, and insects.

A small quantity of wet grain may be dried by putting it into a box that is enclosed with screen wire on all six sides. Ordinary house screen wire should be reinforced with half-inch mesh wire to support the weight of the grain. The wet grain may also be poured into bags and placed on end on slats in a dry place, preferably not on a concrete or earth floor. After a few days the boxes or bags should be inverted and the grain disturbed to permit air to get to all the kernels. Invert again in a week or so and allow it to cure at least one month before attempting to store or mill it.

Keeping Rats Out of Your Grain

Dry grain keeps best when stored in a cool, dry place. It may be stored in bags on slats, so long as precautions are taken against rat infestation. Rats are capable of doing severe damage to stored foods. Not only can they consume many pounds of food in one year, they can also contaminate foods by carrying into the storage area insects and bacteria. You'll know you have a rat problem when you see droppings, tracks, and grease marks. In dusty locations you may notice tracks left by the trail of a tail and the four-toed front paws and the five-toed rear paws. Storing bags of grains in a cellar or outbuilding that has double walls and space between floors and ceilings invites rat problems because such structures provide ideal homes for these rodents. Choose a storage area that is rat-tight and provides no hidden areas in which rats can nest.

Many commercial warehouses take care of rat problems by using rodenticides. These poisons are not only toxic to rats, but to domestic animals and humans as well. For the homesteader, rat infestation can be prevented by making sure the storage area is clean and rodent-proof. Having a couple of good-working felines around to patrol the area is also a good idea. Perhaps the simplest and most logical way to protect your grain is to store it in metal drums in a dry place, not in the cellar, if possible, although this is all right, too, if the cellar is dry and the drums are kept a couple of feet above the floor in case of flooding. Small quantities of cured grains may be kept in glass jars; light helps to inhibit mold.

Preventing Insect Infestation

Weevils, moths, beetles, and the other insects that feed on grain are unlikely to cause you any trouble if your grains and storage containers are clean and dry and kept relatively cool. If insects should invade your grain, remove and destroy that grain which is infested. If, after examining the grain, you find that little damage has been done, you may kill the insects by putting the grain on trays and heating it to 140 to 160° F. in the oven. Or you may put your grain in the freezer and leave it there for about twenty-four hours. Such temperature extremes will destroy insects. Carefully clean out your storage containers, making sure that you have not overlooked any place where insects may be hiding. Pour the cleaned grain back into the containers and check frequently for future infestations.

You can keep unmilled grain in glass jars for up to one year. Make sure that the grain is dry and free from insects before storing it this way.

Refrigerated Storage

At Walnut Acres, organic grower Paul Keene keeps all his grains under refrigeration. This not only assures him of fresh grains, but also makes insect and rodent damage nonexistent. Once, when he spotted some grain moths flying about in the refrigerator as grain was coming in from the weeding out process, he became worried. But the moths settled on the refrigerator wall, only to fall dead on the floor. The 40° F. temperature did it.

Refrigeration of grains is a good idea, especially if you want to hold grain over the summer, when the cool weather is gone. Insects that attack grains multiply at temperatures greater than 70° F. Insect infestations of this type are controlled at the commercial level with fumigants like methyl bromide and methyl parathion, both deadly poisons that are pumped into airtight sealed bins. Experiments are being done with carbon dioxide fumigation, which, while deadly to insects and rodents, will leave no toxic residues. But, like all fumigants, it requires sealed storage bins, and these are not very practical for the grower with small grain acreage because they are usually too big and too expensive to construct. The same insects that multiply at 70° F. lie dormant at 35° F. and will eventually be killed by extended periods of low temperatures. Refrigeration also aids in controlling fungus attack: fungus grows very slowly at temperatures of 45° F. and below.

Organically Grown Grains Store Better

Organic growers may be happy to learn that Paul Hawkin, president of Erewhon Trading Company, which is one of the country's largest distributors of organic grains, claims that organic grains have a higher resistance to heat and humidity than chemically grown grains. "For example." says Paul, "if you fill a pint jar half full with organically grown and produced rice, fill it with water and cap it tightly and then do the same thing with the chemically grown rice, you'll discover a strange thing. After a period of several months the rice that was not grown organically will have dissolved and turned into a white powder, whereas the organically grown rice has not. Now, I have no scientific explanation for this, but it is pretty graphic. I spent a year in Japan studying natural foods and farming methods, and one of the most stunning demonstrations I saw was this rice experiment done by the Messian people in Shikoku. They performed dozens of similar experiments, using other grains as well."

Grinding Grains

Grinding for flour is best done just before baking, for grinding will expose the germ which will turn rancid in the presence of oxygen and warm temperatures. Rancidity destroys vitamins E, A, and K in the human body and has been found to destroy several of the B vitamins. Ground flour should be stored in the refrigerator or freezer if kept for any extended length of time. However, do not use cold flour when you're baking with yeast. Leave the cold flour at room temperature for a few hours before you use it for baking so that the yeast will not be chilled and become inactivated.

STORING BREAD

Before the time of factory-made bread, when all bread was made at home, one day a week was set aside as bread baking day. On that day, the baker of the household would rise before dawn and build a roaring fire in the brick oven. By mid-morning, when the oven's inner walls were thoroughly heated, the glowing embers would be shoveled out. Then the bread, which had been mixed, kneaded, left to rise, and shaped into loaves, and left to rise again, would be placed in the oven. The heavy iron oven door would be closed and sealed tightly with clay so that no heat could escape, and the bread be left to bake for the rest of the morning and afternoon. The farmhouse soon would be filled with the sweet yeasty smell of baking bread, and by early evening there would be enough loaves cooling in the kitchen to last a family of ten or twelve for a week, until the next bread baking day came around again.

It made good sense then—and still does today—to bake a number of loaves at one time. Although we're spared the chore of building a fire in the brick oven, we still must mix and knead the dough, let it rise twice, and bake it. It takes just about as long to prepare and bake a double or triple bread batter as it does a single one. Once you have the kneading board out, the bread pans lined up, and your hands covered with flour, you might as well go ahead and bake enough bread to last your family at least a week.

What do you do with the eight or twelve loaves after they're baked? Farmwives used to put them in a wooden box especially made for keeping bread, cover them with a clean towel, and store them in a cool place. The bread tasted just great for the first few days, but as the week wore on the bread started to get a little stale—no preservatives to keep bread "tasting" fresh forever, then.

Freezing Baked Bread

Today we're more fortunate. We have a way of keeping bread fresh for months, not by adding sodium or calcium propionate or another bread freshener to our breads, but by using the freezer.

Baked yeast breads retain their just-baked quality at freezer temperatures for six months and more if they are stored properly. Prepare and bake your loaves according to the recipe. Allow them to cool in a draft-free place (drafts tend to shrink baked goods) thoroughly before wrapping for the freezer. Warm or hot breads which are wrapped tightly will emit water vapor while they cool and this water vapor will condense on the inside of their wrappings. This moisture can lead to the growth of mold even under the most sanitary of conditions. Wrap the breads intended for the freezer after they are cool in aluminum foil or heavy-duty plastic bags. Then freeze them.

Thawing Your Bread

Frozen breads should be thawed while still in their wrappers. If you like to thaw your breads in a low oven, your freezer wrap should be aluminum foil. Otherwise, let your bread thaw at room temperature. It should take at least an hour to thaw normal-sized loaves. Allow extra thawing time for large, heavy loaves. If you're toasting your bread, there's no reason to thaw it first. Slice it (you'll need a good sharp knife to get through a loaf of bread which is frozen solid) and place it still frozen in the toaster. The slices will thaw and toast at the same time and taste just as good as fresh toasted bread.

Start With Moist Breads

Bread should be eaten soon after it has thawed, for thawed bread dries out quickly. To keep your bread from becoming stale too rapidly after thawing, begin with a recipe that makes a moist bread. One that calls for honey will make a moister loaf than one that uses sugar, corn syrup, cane syrup, maple sugar, or molasses as a sweetener. The more honey you use, the moister your bread will be and the longer it will keep. Honey is hygroscopic. It absorbs moisture from the air and holds this moisture in bread. Recipes that call for a good amount of oil or butter will also make a moist bread that will keep for a longer time because shortening doesn't dry out. Don't freeze French and Italian breads unless you know that you're going to finish off the entire loaf soon after it is thawed, or unless you have a use for stale, dried-out bread. French and Italian breads do not contain oil or butter, so they dry out very quickly.

Foods with a high oil content, like nuts, sunflower and sesame seeds, and soybeans, can be ground in a grinder or blender and used to replace some of the flour in your bread. In addition to making your bread moister, these foods will make your bread more nutritious because they are good sources of protein and the B vitamins. You can replace about one-eighth of the flour in any dark bread recipe with ground nuts, seeds, or soybeans without making an appreciable change in the taste or texture of the finished product. In place of the ground nuts or seeds, you can add a few tablespoons of a nut or seed butter (like peanut butter or tahini) to your bread ingredients. Another way to make your bread moister is to replace some of the flour in the recipe with cooked cereal, like oatmeal or farina.

Softening Dried-Out Bread

If your thawed bread starts to dry out and get stale, do what the farmwives used to do with their six-day-old bread: rejuvnate it by sprinkling a little water on it and putting it in the oven for a few minutes. You can also lay a flat strainer above a pot of boiling water, or cooking vegetables, or soup, or anything you've got on the stove, and place slices of bread in the strainer. Put the lid over the bread. The steam rising from the pot below will moisten and warm your dry pieces of bread.

Freezing Unbaked Bread Dough

If you're the kind of person who goes crazy over the tastes and smells of bread right from the oven and would rather take the time to bake every day just so that you can enjoy just-baked bread all the time, don't despair; there is a short-cut for you, too. Just freeze your bread *before* it's baked, and then when you're ready for a loaf, let it thaw and rise, and bake it as you would freshly risen dough. Your kitchen can smell like it's bread-baking day every day even though you only get your hands and the kneading board floury once a week.

Unbaked yeast dough can only be stored in the freezer for two to three weeks. Mix and knead it the way you usually do, then allow it to rise once. When it's risen to double its bulk, punch it down, shape it into loaves, and freeze it. Your loaves should be thinner than usual. They should be no more than two inches deep so that they will thaw quickly when taken from the freezer. When you're ready for a fresh loaf, thaw the dough in a low oven (250° F.) for forty-five minutes, then reset the oven for the normal baking temperature and bake as usual. Once cooked, this bread will also dry out quickly, so eat it soon after baking.

Refrigerating Unbaked Bread Dough

Floss and Stan Dworkin, in their book, *Bake Your Own Bread* (Holt, Rinehart and Winston, 1972), recommend storing unbaked bread dough in the refrigerator instead of the freezer. It keeps just as long at 40° F. as it does at 0° F.—two to three weeks. In the refrigerator you can get the dough to rise an extra time because the yeast is still working at refrigerator temperatures, even though it does work slower when it is chilled. This extra rise can only improve the texture of the bread.

To prepare the dough for keeping in the refrigerator, let it rise once until double in bulk, and then punch it down in the bowl. Cover the bowl loosely with plastic wrap or with a clean towel and put it in the refrigerator. If the dough rises above the rim of the bowl, punch it down and cover it again. When you're ready to bake the bread, take the dough from the refrigerator and punch it down. Shape it into loaves and put it into bread pans. Place the pans in a warm place and let the dough rise until it is double in bulk. Bake the bread as normal.

Oatmeal Bread

- 1 package yeast
- 2 cups warm milk
- ½ cup honey
- ½ cup oil
- 1 beaten egg
- 1 teaspoon salt
- 4 cups wholewheat flour (or favorite combination of flours)
- 2 cups rolled oats or oat flour

Sprinkle the yeast into the warm milk, add the honey, Mix these ingredients together well, then add the oil, egg, and salt. Stir in the wholewheat flour and add the oat flour so that the dough is dry enough to leave the sides of the bowl.

Turn out onto a floured board and knead until dough is smooth and elastic, about 10 minutes. Place the dough in a large, oiled bowl; cover and allow to rise until double in bulk.

Punch down and cut into 3 pieces. Shape into loaves and place in bread pans. Let rise until double in bulk, then bake at 375° F. for about 45 minutes. Yield: 3 loaves

Peanut Bread

2 teaspoons dry yeast
1¾ cups warm water
3 tablespoons honey
1½ teaspoons salt
¼ cups milk
1½ cups peanut flour
4½ cups wholewheat flour (or favorite combination of flours)

Sprinkle yeast over the water. When the yeast is dissolved, add the honey, salt, milk powder, and peanut flour. Add enough wholewheat flour to form a moderately stiff dough.

Turn the dough out onto a wooden board and knead until smooth and elastic, about 7 minutes. Place dough in a large, oiled bowl, cover and allow to rise until double in bulk.

Punch down dough, shape into one loaf and place in a large oiled bread pan. Cover and allow to rise until double in bulk.

Bake at 350° F. for about 50 minutes. Yield: 1 large loaf

Crusty Loaves

12½ to 13½ cups flour (unbleached, wholewheat, or a combination of your favorites)
½ cup honey
2 tablespoons salt
⅔ cup instant nonfat dry milk solids
4 packages active dry yeast
¼ cup oil
4 cups very warm tap water (120°F. – 130° F.)

In a large bowl thoroughly mix 4 cups flour, honey, salt, dry milk solids and undissolved active dry yeast. Add oil.

Gradually add very warm tap water to dry ingredients and mix well. Stir in enough additional flour to make a stiff dough. Turn out onto lightly floured board; knead until smooth and elastic, about 15 minutes. Cover with a towel; let rest 15 minutes.

Divide dough into 4 equal pieces. Form each piece into a smooth ball. Flatten each into a mound 6 inches in diameter. Place on oiled baking sheets. Cover sheets tightly with plastic wrap. Freeze until firm. Transfer to plastic bags. Keep frozen up to 4 weeks.

Remove from freezer; unwrap and place on ungreased baking sheets. Cover; let stand at room temperature until fully thawed, about 4 hours. Roll each mound to a 12 x 8-inch rectangle. Beginning at an 8-inch end, roll dough as for jelly roll. Pinch seam to seal. With seam side down, press down ends with heel of hand. Fold underneath. Place each, seam side down, in an oiled 8½ x 4½ x 2½-inch loaf pan. Cover; let rise in warm place, free from draft, until doubled in bulk, about 1½ hours.

Bake at 350° F. 30 to 35 minutes, or until done. Remove from pans and cool on wire racks.

To make round loaves: Let thawed mounds rise in warm place, free from draft, until doubled in bulk, about 1 hour. Bake as for loaves. Yield: 4 loaves

Rye Bread

3 packages (or tablespoons) dry yeast
½ cup lukewarm water
1½ tablespoons sea salt
2 tablespoons caraway seeds
2 cups lukewarm water
½ cup oil (safflower, soy or corn)
5 cups rye flour
2 cups rye flour

Sprinkle yeast on top of ½ cup lukewarm water and set aside to dissolve (about 5 minutes).

Combine the salt, caraway seeds, 2 cups lukewarm water and stir in the dissolved yeast mixture. Add the oil and 5 cups of rye flour. Let this "sponge" work for about 2 or 2½ hours in a warm spot in your kitchen. Cover bowl with a damp towel.

Add 2 cups rye flour to "sponge" and turn out of bowl, kneading flour into dough. Place smooth ball of dough in a bowl, cover with damp towel and leave to rise in a warm oven, or some warm spot (about 185° F.). If kitchen is too cool, turn oven on low, briefly, then turn oven off. Set dough on middle rack, placing a pan of water on lower rack to help give moisture.

When dough has grown about ⅓ bigger than it was (after 2 hours or so), knead it again briefly; form it into two round loaves, and place the loaves on a floured cookie sheet and cover with a damp towel. Leave in a warm oven or equally warm spot until the loaves have again grown about ⅓ larger than they were (about 3 – 4 hours). Gently pierce entire surface of loaves with prongs of a fork.

Bake in a 375° F. oven for about 30 minutes. Cool loaves on rack before cutting. Yield: 2 loaves

Cream, Honey and Wheat Bread

- 1 cup very hot water
- ⅓ cup honey (more if desired)
- 1 cup heavy sweet cream
- 2 packages yeast
- 4 large eggs
- 5 cups wholewheat flour

Dissolve honey in water and add cream. Dissolve yeast in this mixture when warm. Add 2 cups wholewheat flour and mix well. Add the eggs and mix. Add rest of flour, mix, then turn out on floured board and knead until smooth.

Let rise until double in bulk, shape into loaves, buns or sweet rolls, let risc and bake in oven about 20 – 30 minutes at 375°. *Do not* let dough rise too much in pan as it will rise in the oven more than ordinary bread. Yield: 2 loaves

These quick breads are very moist and make good freezers. They stay fresh and moist many days after they are thawed. Freeze them after they have been baked and cooled.

Pumpkin Bread

- 4 beaten eggs
- 2 cups cooked pumpkin, fresh or canned
- ⅔ cup water
- 1 cup oil
- 2 cups honey
- 2 teaspoons baking soda
- 1½ teaspoons salt
- 1 teaspoon cinnamon
- 1 teaspoon nutmeg
- 3½ cups unbleached flour

Beat all ingredients together well. Pour the mixture into two large, oiled bread pans. Bake 1 hour and 10 minutes at 350° F. Yield: 2 loaves

Carrot Corn Bread

- 1 cup yellow corn meal
- 1 cup grated carrots
- 2 tablespoons vegetable oil
- 1 tablespoon honey
- 2 eggs, separated
- ¾ cup boiling water
- 2 tablespoons cold water

Mix thoroughly corn meal, carrots, oil, and honey. Stir in the boiling water. Add 2 tablespoons water to the 2 egg yolks beaten together. Add this to the above mixture. Fold in stiffly beaten egg whites. Pour into a warm oiled pan 8 x 8, and bake at 400° F. for 25 minutes, or until knife comes out clean. Yield: 6 servings

Honey Applesauce Oatmeal Bread

½ cup honey
½ cup oil
1 cup canned applesauce
⅜ cup lukewarm milk
1 tablespoon honey
2 packages active dry yeast
2 eggs
3 cups sifted unbleached flour
1 cup rolled oats
1½ teaspoons salt
applesauce topping
Cinnamon
Nutmeg

Combine honey, oil, and applesauce. Heat until shortening melts. Cool to lukewarm. Combine milk, 1 tablespoon honey and yeast, stirring until yeast dissolves. Let stand 5 to 10 minutes.

Beat eggs in large bowl. Add lukewarm applesauce mixture, yeast mixture and flour. Mix to smooth batter. Add oats and salt; mix well. Cover and let rise until double in bulk. Beat batter again, spread batter in a greased 8″ round spring form pan.

Spread Topping on dough. (See below for recipe.) Sprinkle with nuts, cinnamon and nutmeg.

Cover and let rise until double in bulk. Bake in 375° F. oven, 50 to 60 minutes or until done. Yield: 1 round loaf

Applesauce Topping

1 cup canned applesauce
2 tablespoons butter
¼ cup honey
½ cup coconut

Slowly cook applesauce down to ½ cup; combine with remaining ingredients.

Millet Bread

1½ cups millet flour
1 cup grated carrots
1 tablespoon honey
1 teaspoon sea kelp
3 tablespoons cooking oil
3 eggs
¾ cup boiling water

Separate the eggs and beat the whites very stiff. Set aside. Pour the boiling water over the millet flour, then add the carrots, honey, kelp, and oil. Beat the egg yolks and add 3 tablespoons cold water. Add to the millet mixture. Fold in the egg whites last, pour the batter in a very hot oiled bread tin and bake about 45 minutes at 350° F. The oven should be preheated while the bread is being stirred up, and the oiled tin put in the oven to heat. This bread must start baking immediately. It tastes and looks like corn bread, but is a richer yellow because of the carrots. Yield: 1 loaf

Banana Corn Bread

- 1¼ cups corn meal
- ¼ cup wheat germ
- ¼ cup potato flour
- ½ cup soybean flour
- 1 tablespoon or package of baking yeast softened in 2 tablespoons of warm water
- 2 eggs well beaten
- ½ cup mashed banana
- 7 tablespoons honey
- 2 tablespoons corn oil

Place all dry ingredients in a bowl. In another bowl, beat the eggs and all other ingredients. Add the dry ingredients to the egg mixture and mix well. Place in an oiled bread pan and let rise about 1 hour. Bake in a 325° F. oven for 50 minutes. Yield: 1 loaf

INDEX

A

B

C

D

E

F

G

H

J

K

L

M

N

O

P

Q

R

S

T

U

V

W

Y

Z